Experiments in General Cl
Laboratory Manua

General Chemistry

TENTH AND ELEVENTH EDITIONS

Darrell D. Ebbing
Wayne State University

Steven D. Gammon

Prepared by

R.A.D. Wentworth
Indiana University (Emeritus)

Barbara H. Munk
Wayne State University

Australia • Brazil • Mexico • Singapore • United Kingdom • United States

WCN: 01-100-101

For product information and technology assistance, contact us at **Cengage Learning Customer & Sales Support, 1-800-354-9706**.

For permission to use material from this text or product, submit all requests online at **www.cengage.com/permissions**
Further permissions questions can be emailed to **permissionrequest@cengage.com**.

ISBN: 978-1-305-94498-5

Cengage Learning
20 Channel Center Street
Boston, MA 02210
USA

Cengage Learning is a leading provider of customized learning solutions with office locations around the globe, including Singapore, the United Kingdom, Australia, Mexico, Brazil, and Japan. Locate your local office at: **www.cengage.com/global**.

Cengage Learning products are represented in Canada by Nelson Education, Ltd.

To learn more about Cengage Learning Solutions, visit **www.cengage.com**.

Purchase any of our products at your local college store or at our preferred online store **www.cengagebrain.com**.

Printed in the United States of America
Print Number: 01 Print Year: 2016

Contents

Preface

Experiments in General Chemistry has been written to accompany the tenth and eleventh editions of *General Chemistry* by Darrell D. Ebbing and Steven D. Gammon. Although this laboratory manual can be used with other textbooks, it uniquely complements the organization and coverage of the Ebbing/Gammon text. As a result, an immediate advantage is realized: The lecture and laboratory parts of a course can be integrated with relative ease.

Organization and Content of the Traditional Experiments

Forty-one traditional experiments in this manual are arranged to parallel the material found in the 24 chapters of the Ebbing/Gammon book. At least one of these experiments is provided for each chapter of the text. For 14 chapters, two or more experiments are offered for each chapter, allowing better coverage of important concepts if time is available, or a choice of topics if it is not. The experiments range from qualitative inspections of various reactions to quantitative interpretations of stoichiometry, analysis, and many chemical phenomena. Four experiments incorporate some aspect of thermochemistry because of heat's importance in experimental chemistry.

Each of the traditional experiments is numbered according to the Ebbing/Gammon chapter with which it can be used. For example, Experiments 1A, 1B, and 1C can be used with Chapter 1, and Experiment 2 is intended for use with Chapter 2. References to relevant sections in the Ebbing/Gammon text that appear throughout this manual enable the student to see and appreciate the close relationship between the lecture and laboratory parts of the course. They do not necessitate the use of the Ebbing/Gammon text, however. A student will not need to refer to the textbook to understand the laboratory manual. All required explanations and data are given in this manual.

Flexibility is a key feature of this manual. In some instances, the traditional experiments may be used appropriately in conjunction with any one of two or three chapters in the textbook. In other cases, alternative experiments are available should the instructor choose not to do a given one. A complete discussion of various sequences of experiments and their alternatives can be found in the *Instructor's Resource Manual* that accompanies this manual; it is available to instructors via the Instructor's Resource Center at **www.cengage.com/chemistry/ebbing/generalchemistry11e**.

Format of the Traditional Experiments

Each of these experiments is carefully organized with introductory remarks, an in-depth discussion of the experiment's purpose, and step-by-step procedures. All experiments include a prelaboratory assignment to be completed by students before coming to the lab. Separate sections are provided for noting the results obtained during the experiment and for questions to be answered after the lab's completion. All three sections—prelaboratory assignments, results, and questions—are on perforated pages that can be detached easily and handed in to the instructor.

Reliability and Time Requirements of the Traditional Experiments

Each of these experiments has been tested and used for years by literally thousands of students. Many of the experiments can be completed in about 2 hours, but a few require as much as another hour. The *Instructor's Resource Manual* lists ways to shorten these longer experiments to about 2 hours.

New Challenging Inquiries with Limited Guidance

Through the previous editions of this manual, the experiments were for the most part traditional—traditional in the way they were presented and traditional in what was expected of the students. This edition retains those experiments, 41 in number, but it now includes ten inquiries with limited guidance, experiments carefully designed to allow students to work at their own intellectual level, design their own experiments, and analyze the data from those experiments without help or prompting from the manual. None of these inquiries is tied necessarily to any chapter of the Ebbing/Gammon text. All can be used whenever the instructor wishes. They are numbered I-1, I-2, and so on.

Safety

Safety issues receive special attention throughout this manual, including the new inquiries with limited guidance. General safety rules are given in the Introduction. In each experiment students are reminded of these rules, and they are also given special precautions formulated for that particular experiment. These are highlighted by the heading "CAUTION." All of these precautions have been thoroughly reviewed. Instructors and students should refer to reagent manufacturer's safety data sheets to ensure students wear appropriate personal protective equipment while running experiments.

Acknowledgments

I want to acknowledge the invaluable contribution of freshmen and teaching assistants at Indiana University. Without their help, I could not have written this manual.

R. A. D. W.

*A note about our colleague, Rupert Wentworth (Indiana University, Emeritus), who passed away on September 5, 2007. A mainstay of the *General Chemistry* team since the program's inception, Rupert authored *Experiments in General Chemistry*. He is sorely missed and fondly remembered.

-Darrell D. Ebbing

-Steven D. Gammon

Introduction

The study of chemistry is a fundamental part of any science curriculum. Because chemistry has developed largely through experiments, the study of chemistry is augmented by laboratory experiences that demonstrate, clarify, and develop still further the principles of chemistry discussed in the classroom. This tested laboratory manual, along with thoughtful discussion with your laboratory instructor, will help you with the first steps toward the realization of these objectives. However, you must shoulder a significant portion of the burden.

The laboratory environment differs considerably from the outside world, and you must accept the abrupt change that you will experience upon entering the laboratory. In a real sense, you must adopt a new lifestyle for a few hours each week. The lifestyle in the laboratory is much more structured than your lifestyle at home, or even in the classroom. Certain codes of conduct pertaining to safety and "housekeeping" must be followed at all times. Moreover, you must adopt correct procedures for using glassware and other pieces of equipment.

Once you learn the following operating rules and adopt them as a code to be followed for a short time each week, you will be able to appreciate your laboratory experiences to a greater degree. You will be able to complete the experiments given in this manual in a relatively safe environment. A proper respect for good housekeeping practices will prevent contamination of the laboratory's chemicals. This, and your correct use of the glassware and other equipment, should ensure your success with these experiments.

Safety Rules

Work in a laboratory should be a safe experience. It will be safe, however, only if certain safety guidelines are followed without exception. Safety is up to you. Before the 19 rules are listed, it is worthwhile to consider their common origin: Each rule is the direct outgrowth of several accidents. Only with hindsight do we tend to ask, "What regulation would have prevented these accidents?" Once each rule has been formulated, it seems so reasonable and so much in accord with common sense that we can only wonder why it was not prescribed before the accidents occurred.
The safety rules that will form a large part of your code of conduct in the laboratory are listed below.

1. *Locate the safety equipment.* Find the eye wash fountains, safety showers, fire extinguishers, fire blankets, first aid kit, and all exits that are to be used in an emergency. Your laboratory instructor will describe the use of the safety equipment.

2. *Protect your eyes.* Wear goggles at all times. Prescription eyeglasses, if you need them, must be worn under goggles. You should not wear contact lenses in the laboratory because various fumes may accumulate under the lenses and injure your eyes.

3. *Tie long hair back.* Observing this precaution will keep your hair out of burner flames and harmful liquids.

4. *Wear shoes that cover all of your feet.* Broken glass on the laboratory floor is all too common.

   Your feet will need more protection than that afforded by open-toed shoes or sandals.

5. *Wear clothes that cover your torso and your legs to the knees.* Clothing will give your body needed protection against accidental contact with chemical reagents. Good clothing can be protected with a laboratory apron or coat.

6. *Do not eat or drink in the laboratory.*

7. *Do not taste any chemical.*

8. *Do not smell chemicals directly.* Use your hand to waft the odor to your nose.
9. *Do not pipet solutions by mouth.* Use a rubber suction bulb or special pipet filler.
10. *Do not put flammable liquids near an open flame.*
11. *Do not engage in games or horseplay in the laboratory.*
12. *Do not do or watch unauthorized experiments.*
13. *Do not work in the laboratory in the absence of your laboratory instructor or his or her authorized representative.*
14. *Use a hood when required.* This manual will indicate when a hood (or another similar device) is needed.
15. *Handle glass tubing with care.* Glass tubing is easily broken. When tubing (including thermometers) is to be inserted through a rubber stopper, the tubing must be lubricated with water or glycerol. Hold the tubing with a cloth or a paper towel near the end that will be inserted, and use a twisting motion during insertion.
16. *Be aware of your neighbors. Are they obeying the safety rules?* A neighbor's accident may not hurt him or her, but it may injure you.
17. *Wash your hands before leaving the laboratory.*
18. *Tell your laboratory instructor about an accident or a spill immediately.*
19. *Do not violate any other safety rule given in this manual or issued by your laboratory instructor.*

Housekeeping Rules

All of you realize that good housekeeping in your home results in a pleasant place in which to live. Good housekeeping in the laboratory will likewise lead to pleasant surroundings. In addition, it will provide a safe work site in which you may be assured that chemicals are not contaminated. Please observe the following rules.

1. *Clean up broken glass immediately with a broom and a dustpan. Do not use your hands.* Special containers may be provided for disposal. Ask your laboratory instructor to answer any questions you have regarding proper disposal of broken glassware.
2. *Clean up solid and liquid spills immediately, but only after checking with your laboratory instructor about possible safety hazards.*
3. *Do not pour any chemical into a sink without authorization from the laboratory instructor.* Often, disposal bottles will be provided. Ask your laboratory instructor for guidance if you are unsure how to properly dispose of excess chemicals.
4. *Take containers to the stock of chemicals.* Do not bring stock chemicals to your laboratory table.
5. *Read the label on a bottle carefully.* Is it the correct chemical? Is it the correct concentration?
6. *Do not insert a pipet or medicine dropper into a stock bottle.* Avoid contamination by pouring the liquid into one of your test tubes, flasks, or small beakers before taking a sample.
7. *Use special care with stoppers or tops of stock bottles.* Do not allow them to pick up contamination. Your laboratory instructor may provide additional instructions for handling the stoppers or tops found in your laboratory.
8. *Take no more of a chemical than the experiment requires. This practice reduces chemical waste and is better for the environment.*
9. *Never return an unused chemical to a stock bottle.* You must assume that the chemical is

contaminated. It must be discarded.

10. *Set up your glassware and apparatus away from the front edge of your laboratory bench.*

11. *Follow any other housekeeping rules given by your laboratory instructor.*

Common Glassware and Other Equipment

The glassware and other equipment that may be found in most general chemistry laboratories is shown in Figure I.1. Refer to this figure whenever an experiment calls for a piece of unfamiliar equipment. The use of certain pieces of this equipment will now be discussed.

Volumetric Glassware

Graduated cylinders, transfer pipets, Mohr pipets, and burets are used to make volumetric measurements. This glassware must be scrupulously clean. Liquids must drain without leaving drops adhering to the inner walls of the glassware.

Graduated Cylinders

A graduated cylinder is used to measure an approximate volume of a liquid. When water or an aqueous solution (a solution containing water) is added, the upper surface of the liquid in the graduated cylinder will be concave. This concave surface is called a *meniscus*. The bottom of the meniscus is used for all measurements. To avoid error (called parallax error), your eye should always be level with the meniscus when you are measuring the volume. Figure I.2 shows the correct way to read the volume in a graduated cylinder..

Some graduated cylinders are calibrated *to contain* the volume that is measured. You will usually find TC etched in the glass of this type of graduated cylinder. The delivered volume will always be slightly less than the measured volume because of the residual liquid that coats the inner walls and remains after the liquid is poured out. Other graduated cylinders are calibrated *to deliver* (TD) a measured volume. The actual volume of a liquid in this type of cylinder is always slightly greater than the measured volume in order to compensate for the residual liquid.

Graduated cylinders come in many sizes, but 10-mL, 25-mL, 50-mL, and 100-mL graduated cylinders are often found in general chemistry laboratories. You will be able to measure any volume (±1 mL in many cases) up to the maximum volume of the graduated cylinder.

FIGURE I.1

Common glassware and equipment.

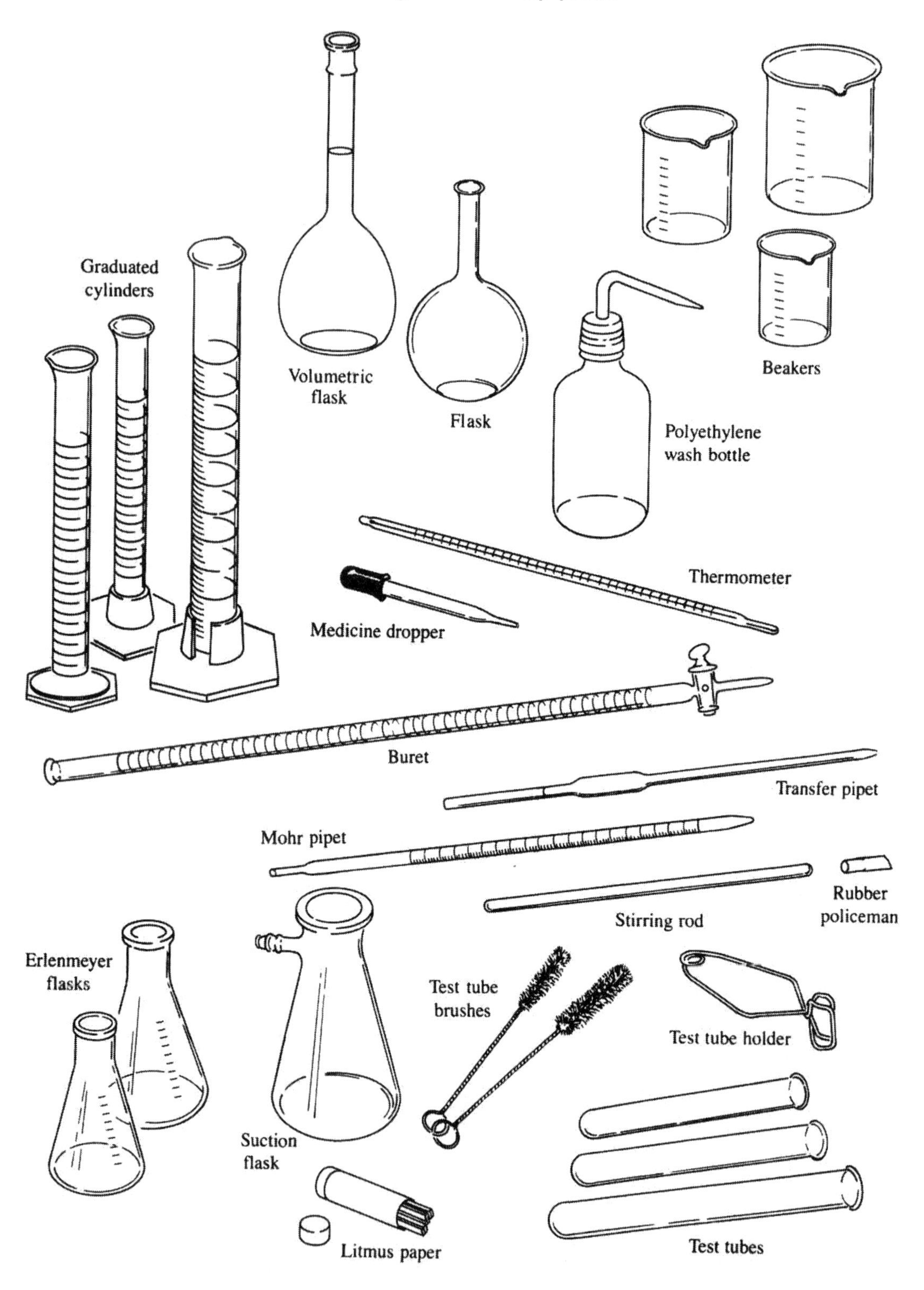

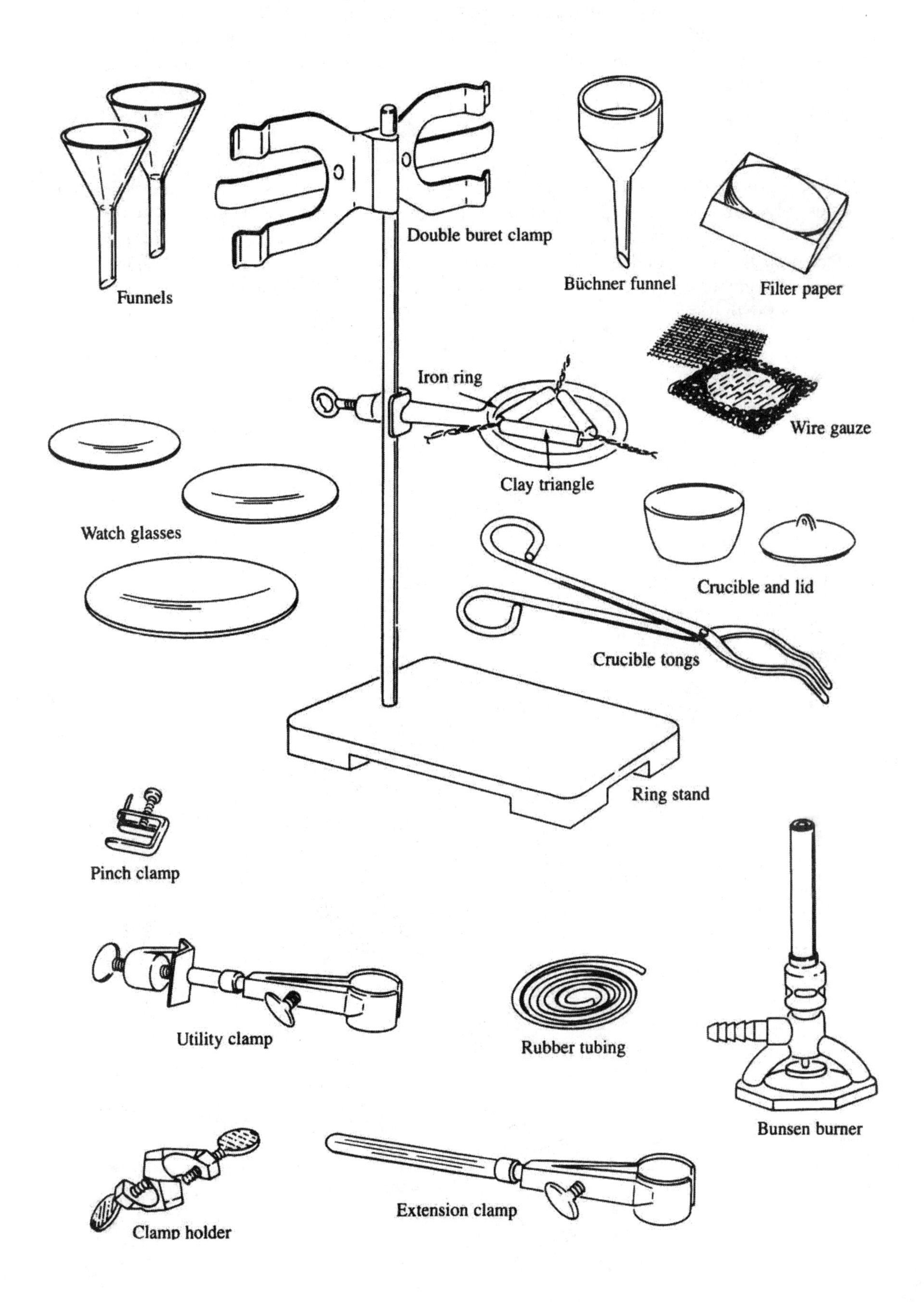
Funnels
Double buret clamp
Büchner funnel
Filter paper
Iron ring
Wire gauze
Clay triangle
Watch glasses
Crucible and lid
Crucible tongs
Ring stand
Pinch clamp
Utility clamp
Rubber tubing
Bunsen burner
Extension clamp
Clamp holder

Pipets

Transfer and Mohr pipets are required for some of the experiments in this laboratory manual. A transfer pipet is calibrated to deliver (TD) one and only one volume, whereas a Mohr pipet is graduated so that it can deliver any volume (usually to the nearest tenth of a milliliter) up to its maximum volume. Transfer pipets come in many sizes, but 5-mL, 10-mL, and 20-mL pipets are usually found in general chemistry laboratories. Mohr pipets are commonly restricted to 5-mL and 10-mL volumes.

The correct use of a pipet requires considerable manipulatory skill. This is not an innate skill but one that comes only with practice. Step-by-step procedures for correct usage with a rubber suction bulb are shown in Figure I.3. If you are to use other suction devices, your laboratory instructor will discuss and demonstrate them. Although the figure shows a transfer pipet, the instructions also apply to a Mohr pipet. Remember that you are not allowed to use your mouth for suction—even if you are filling the pipet with water!

FIGURE I.2

The proper method of reading a meniscus so that parallax error is avoided.

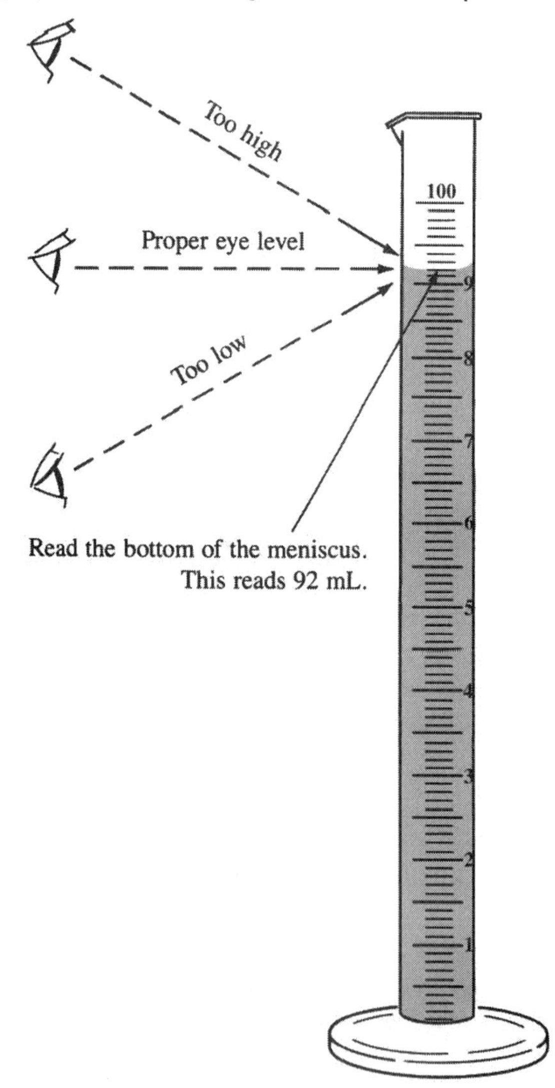

FIGURE I.3

The correct procedure for using a pipet with a rubber suction bulb.

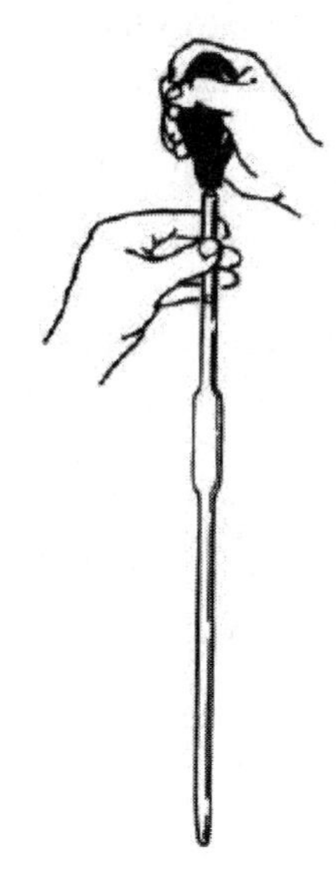

(*a*) Place the rubber suction bulb at the mouth of the pipet, using your more dexterous hand to hold the bulb. Do not insert the pipet into the bulb. Squeeze the air from the bulb.

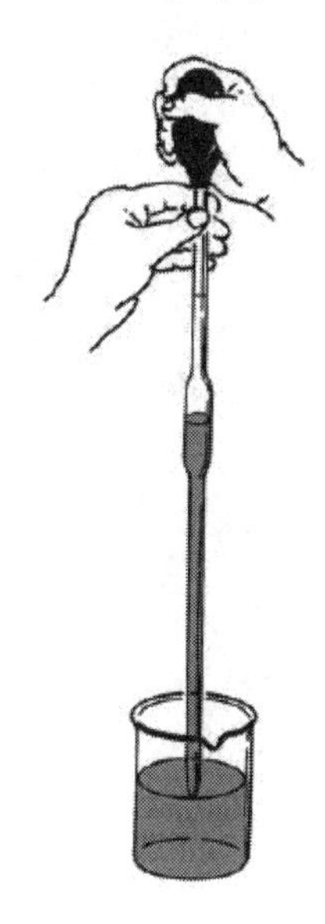

(*b*) Insert the pipet into the liquid. Allow the bulb to expand *slowly* to draw liquid into the pipet. Do not allow the liquid to enter the bulb, where it would be contaminated.

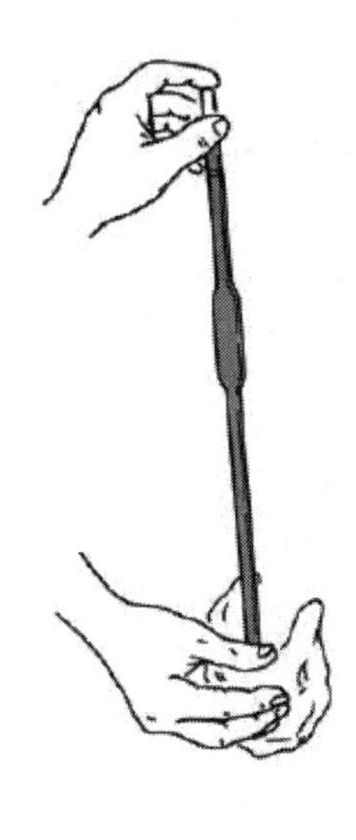

(*c*) When the liquid is about 1 cm above the etched line on the pipet, remove the bulb and place the tip of your index finger from your less dexterous hand over the mouth of the pipet. Remove the pipet from the liquid, and dry its exterior with tissue paper.

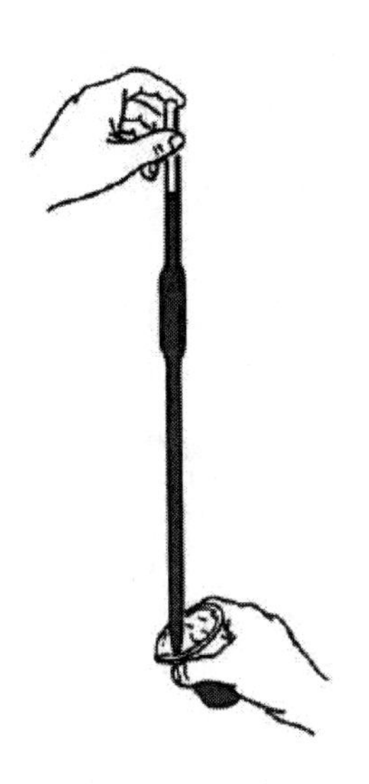

(*d*) Drain the excess into a waste container until the bottom of the meniscus coincides with the etched line. Touch off any adhering last drop.

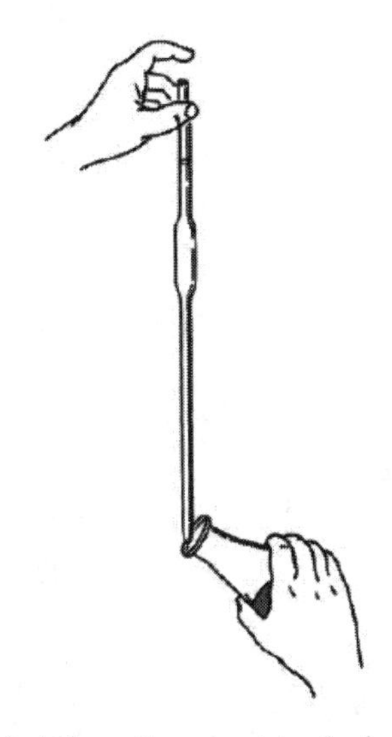

(*e*) Allow the pipet to drain into the container to be used in the experiment. Again, touch off any last drop.

(*f*) Do not blow the remaining liquid from the pipet. The pipet was calibrated to deliver the correct volume with this liquid remaining in it.

Burets

The principal use of the buret is for titrations. Precise titrations require burets that drain freely, are very clean, and do not leak around the stopcock. The following three steps will help you to have a buret that operates as it should.

1. The capillary tip of the buret should be clean and free of foreign objects. A thin wire can sometimes be used successfully to dislodge grease or dirt that impairs or prevents draining.
2. If water droplets are left on the inner walls of the buret after draining, the buret needs a thorough cleaning. It should be cleaned with hot water, detergent, and a buret brush; then it should be rinsed with tap water. Finally, it should be rinsed with distilled water.
3. Some maintenance is required if the stopcock leaks while the buret is draining or if drops form on the capillary tip when the stopcock is turned off. Glass stopcocks must be lubricated to prevent both kinds of leaking. Lubrication will also allow the stopcock to turn easily. Your laboratory instructor will show you how to grease the stopcock without allowing the grease to enter the tip of the buret. Teflon stopcocks do not require lubrication. Leaking can usually be prevented by tightening the tension nut, which seats the stopcock more firmly.

The clean, properly operating buret should now be held in place by a clamp, preferably a buret clamp, which is attached to a ring stand. Before you fill the buret, you should rinse it several times with the solution that will eventually be in it. Place a funnel in the top of the buret, and pour about 3 to 5 mL of the solution through the funnel into the buret. Remove the funnel and take the buret from the clamp. Carefully tip the buret on its side, while holding it with your hand. Do not allow the solution to spill, but tip the buret until the solution comes in contact with almost the entire length. Rotate the buret in your hand so that the inner walls are rinsed completely with the solution. Drain the buret through the stopcock, discard this portion of the solution, and repeat the entire rinsing procedure two more times.

If you are using the buret for the first time, examine its markings before you fill it. The lines that span the entire circumference occur for each milliliter, starting with zero at the top and reaching the maximum volume at the bottom of the buret. As a consequence, the buret will show the volume of a liquid that has been delivered, rather than the volume that remains. The smaller lines indicate each tenth of a milliliter. The spacings between these lines will allow you to estimate the volume to the nearest 0.01 mL. Thus 9.34 mL and 17.60 mL would be typical buret readings. Readings such as 9.3 mL and 17.6 mL are not acceptable because your results will be less accurate if you do not record this important information.

Fill the buret to above the zero mark with the stopcock closed. Open the stopcock fully so that the liquid drains rapidly to flush out air bubbles in the tip of the buret. Drain the buret until the meniscus rests between the zero and 1-mL marks. Do not waste time trying to align the bottom of the meniscus with the zero mark. Read the buret with your eye on the same level as the meniscus. To obtain the volume of the liquid that you use in a titration, subtract this reading from the final reading.

Figure I.4 shows the best technique for a titration by a right-handed student. Note that the left hand is used to open and close the stopcock. Pressure on the stopcock with this hand will keep the stopcock properly seated, preventing leaks. With a bit of practice, you will be able to adjust the stopcock so that as little as half a drop will form on the capillary tip. The right hand is used to swirl the flask.

A left-handed student may turn the buret 180° and then open and close the stopcock with the right hand. Unfortunately, the markings on the buret will now be away from the student and will not be as easy to observe.

FIGURE I.4

Technique for a titration by a right-handed student.

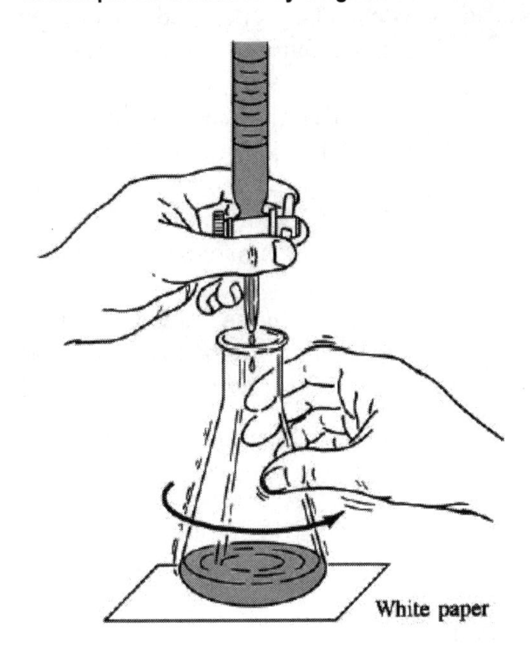

Laboratory Burners

Your laboratory will probably be equipped with Bunsen burners like the one shown in Figure I.5. If you have another type of burner, your laboratory instructor will describe its use.

FIGURE I.5

A Bunsen burner.

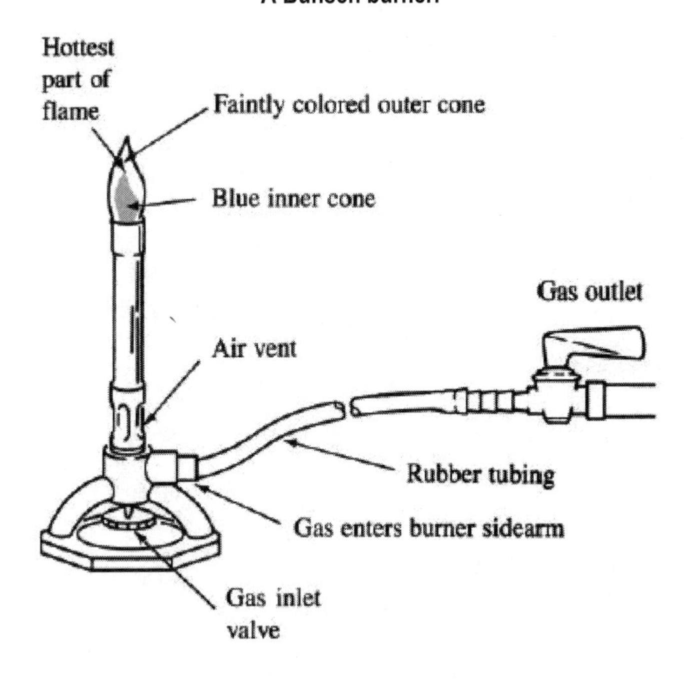

Operating your Bunsen Burner

Your Bunsen burner relies on the combustion of natural gas or bottled gas. To achieve the best flame, the gas inlet valve and the air vents must be properly adjusted. If you adopt the following procedure, you should be able to obtain the best flame for your burner without difficulty.

Your burner should be equipped with a short piece of rubber tubing. One end of the tubing should be attached to the sidearm of the burner, and the other end should be connected to an outlet from a source of gas. Open the gas outlet completely. Strike a match and hold it away from the burner. Open the gas inlet valve on the burner about halfway. Light the burner by bringing the match from the side to the top of the burner. With any other approach, the match may be extinguished by the flow of gas.

Close the air vents and adjust the gas inlet valve until the flame is about 4 inches high. The flame should be luminous and yellow. Open the air vents until the flame becomes two concentric cones. The outer cone will be only faintly colored, but the inner cone will be blue. The hottest part of the flame is at the tip of the blue cone.

Filtration

Filtration through special paper (called filter paper) is the simplest method of separating a solid from a liquid. Filter paper is available with a variety of porosities. A finely porous paper should be used for solids with very small particles, but filtration will be slow. A coarser, more porous paper can be used with solids whose particles are larger. Filtration will then be more rapid.

FIGURE I.6

The correct method of folding filter paper for gravity filtration.

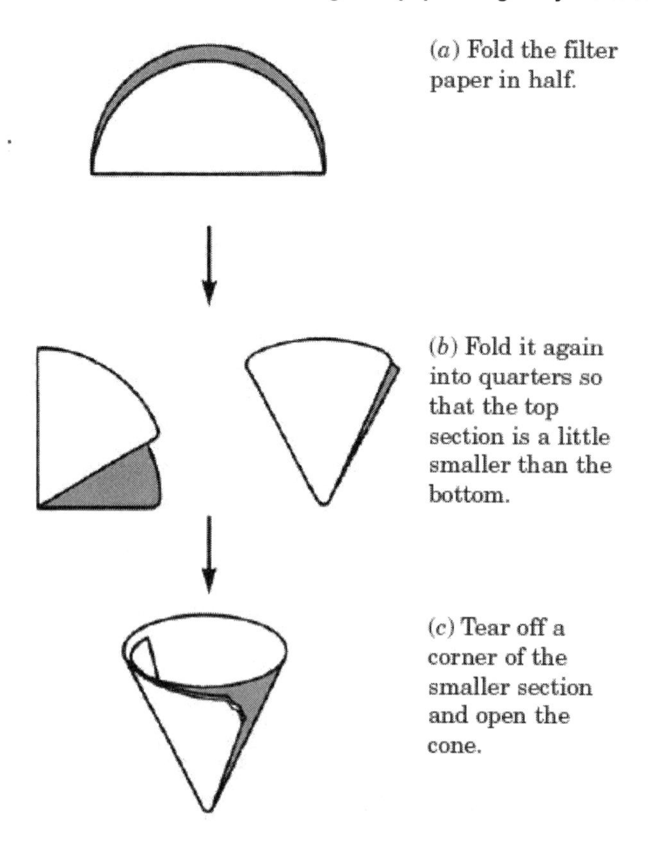

Gravity Filtration

This technique requires a conical filter funnel with a hollow stem and a glass stirring rod. The stirring rod may be equipped with a rubber policeman. This device, which slips over the end of the stirring rod, is used to remove precipitates from the walls of a beaker or flask.

For this method of filtration, the filter paper is folded in the manner shown in Figure I.6. The small tear in the corner of the outside fold permits a closer seal with the funnel during filtration. After you make this paper cone, place it in the filter funnel and wet the paper thoroughly with distilled water. Pour off the excess water and place the funnel in its support. The support can be an iron ring, a clay triangle on an iron ring, a wooden board with a circular hole, or the mouth of an Erlenmeyer flask. If you are filtering into a beaker, you can minimize splashing by putting the stem of the filter funnel against the inner wall of the beaker.

The next step is shown in Figure I.7. The mixture to be filtered should be poured (decanted) along the stirring rod to direct the flow of the liquid into the filter paper. Never fill the filter paper to more than two-thirds of its volume. Most of the solid should be transferred to the filter during this stage. If you have a rubber policeman, use it to remove a solid that adheres stubbornly to the walls of the container.

When all of the liquid has been transferred to the filter, use a stream of distilled water from a plastic wash bottle to rinse the remainder of the solid into the paper. This technique is shown in Figure I.8. When all of the solid has been transferred, rinse the stirring rod in such a way that the distilled water is also directed onto the paper. Finally, wash the solid with two small portions of distilled water. You have now quantitatively transfered the solid.

FIGURE I.7

The correct method of decanting into the filter paper.

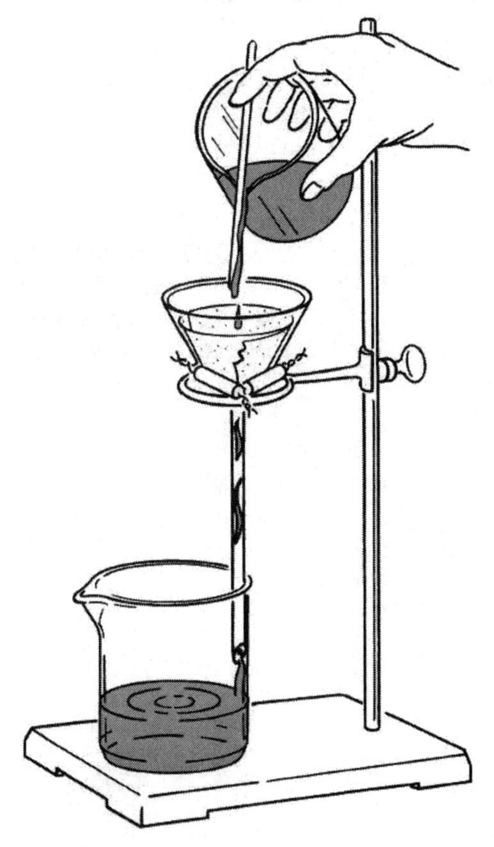

FIGURE I.8

Using a stream of distilled water from a plastic wash bottle to transfer all of the solid into the filter paper.

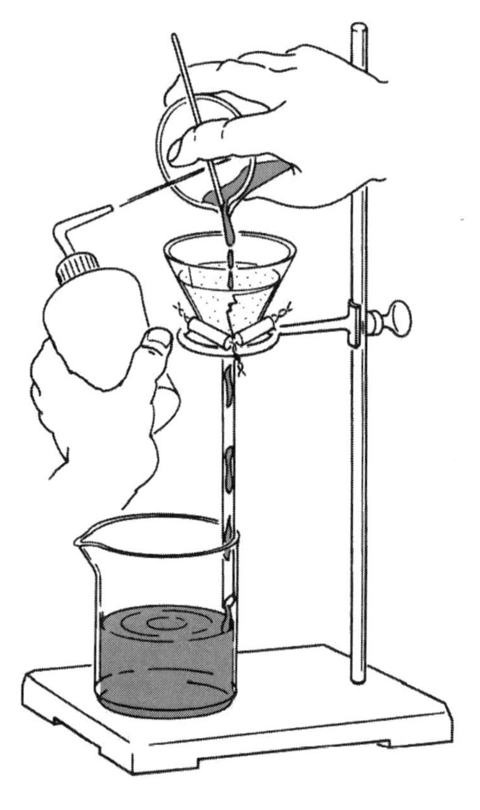

Suction Filtration

This type of filtration is much faster than gravity filtration, but quantitative recovery of a solid is rarely achieved. Suction filtration requires a Büchner funnel, a suction flask, a rubber stopper or rubber ring to hold the funnel tightly in the flask, a glass stirring rod, heavy rubber tubing, and a water aspirator. In addition, it is advisable to use some type of safety trap to prevent water from the aspirator from backing up into the suction flask.

The Büchner funnel and suction flask should be set up as shown in Figure I.9; the flask should be clamped to a ring stand. Place a piece of filter paper of the correct size in the funnel. Turn on the water aspirator to start the suction. Wet the filter paper with distilled water, and allow the water to drain into the suction flask; then empty the flask after all of the water has drained. Place the funnel on the flask once again.

Transfer into the funnel as much of the solid and liquid to be filtered as you can. Use the stirring rod to help you. Repeat this operation until all of the liquid has passed through the funnel. Transfer the filtered liquid back to its original container, and repeat the operation. Eventually, all of the solid will be transferred into the funnel. The solid should be washed with distilled water, or any other liquid that is required, and sucked dry.

FIGURE I.9

The correct setup for suction filtration.

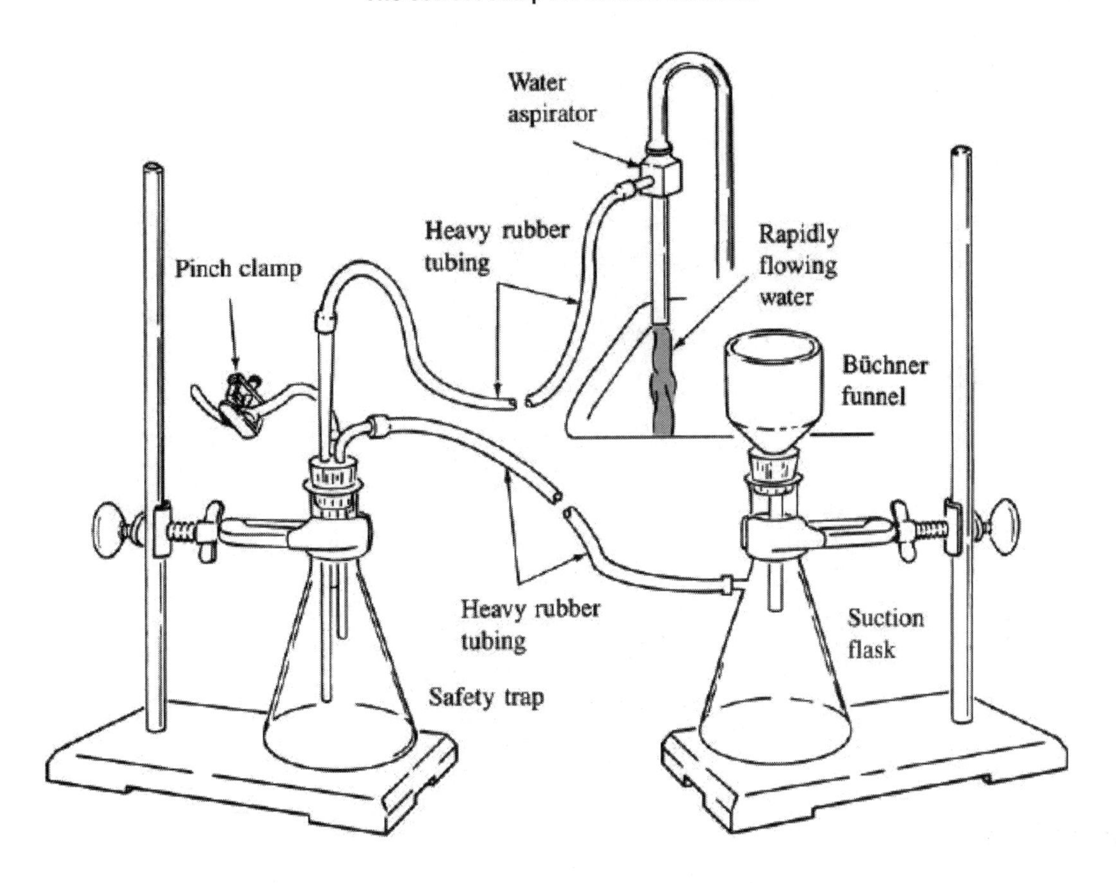

1A. Identification of an Unknown Compound

Introduction

Chemistry is a science that is built on the interrelationship of experiment and theory (Ebbing/Gammon, Section 1.2). Experiments have led to theories, and theories have, in turn, led to other experiments. Accurate and complete observations are required in these experiments to provide a maximum amount of useful information. Without good observations, the cyclic relationships between experiment and theory would be seriously marred and perhaps destroyed.

When a student observes an event in the laboratory, it is not necessarily true that he or she will record a complete and accurate description of that event. Good observations require practice and attention to detail.

Purpose

This experiment emphasizes the importance of observations and the inferences that can be drawn from those observations. You will be able to identify an unknown compound by comparing its reactions with those of some known compounds. Accurate, complete observations are the only requirement for success. No knowledge of the chemistry that you will observe is necessary, and none is assumed.

Concept of the Experiment

You will obtain solid samples of sodium chloride (NaCl), sodium iodide (NaI), sodium carbonate (Na_2CO_3), sodium hydrogen phosphate (Na_2HPO_4), and sodium sulfate (Na_2SO_4), as well as solutions of these compounds in water. You will test these compounds with solutions of nitric acid (HNO_3), barium nitrate [$Ba(NO_3)_2$], silver nitrate ($AgNO_3$), and an acid–base indicator called thymol blue. Simultaneously, you will test an unknown solid sample that is identical to one of the known solid samples. You will be able to determine its identity by matching its characteristic reactions with those of
the known compounds.

Three Important Reaction Signals

In this experiment, certain signals will indicate that chemical reactions have taken place:

1. The color of a solution changes.
2. A gas evolves from the solution.
3. A precipitate appears or disappears.

The first of these signals is easy to interpret and use, but the other two signals require a little more explanation.

When gases are evolved from a solution, you will see bubbles form and move upward through the solution. If you see only two or three bubbles, gas evolution has not occurred. When a gas is evolved during this experiment, it will be very noticeable.

Precipitates are solids, but they may be so finely divided that they appear milky. An example of the formation of a precipitate is shown in Figure 1A.1. Gradually, this precipitate will settle to the bottom of the test tube.

Procedure

Getting Started

1. Obtain the unknown compound from your laboratory instructor.
2. You will also need six test tubes, a medicine dropper, and red and blue litmus paper.
3. Obtain directions from your laboratory instructor for discarding the solutions that you will use in this experiment.

Testing for Gas Evolution

1. Mark your test tubes for recognition with a marking pencil.
2. Use a clean spatula to place pea-sized solid samples of $NaCl$, NaI, Na_2CO_3, Na_2HPO_4, Na_2SO_4, and your unknown in the test tubes.
3. Add 5 drops of the solution of HNO_3 to each test tube, and record your observations.

CAUTION: Nitric acid can cause chemical burns, in addition to ruining your clothing. If you spill any of this solution on you, wash the contaminated area thoroughly with tap water and report the incident to your laboratory instructor. You may require further treatment.

FIGURE 1A.1

The formation of a cloudy precipitate when one solution is added to another.

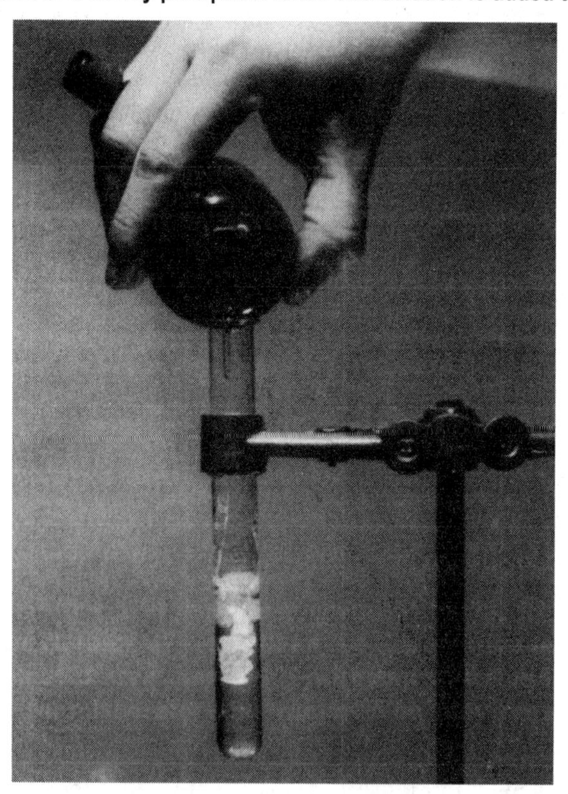

4. Discard the solutions in the test tubes.
5. Wash the test tubes, and rinse them with distilled water.

Dissolving Your Unknown Compound

1. Wash a 100-mL graduated cylinder and a 400-mL beaker thoroughly, and rinse them with distilled water.
2. Place another pea-sized portion of your unknown compound into the beaker. Add 200 mL of distilled water from the graduated cylinder, and swirl or stir gently until all the solid has dissolved.
3. Set aside the remaining portion of your unknown compound in a safe place for use in the event of an unforeseen accident.

Testing with Barium Nitrate

1. Use the solutions of NaCl, NaI, Na_2CO_3, Na_2HPO_4, and Na_2SO_4 that you will find in the laboratory and the solution of your unknown compound that you have just prepared for the remaining tests in this experiment.
2. Place 20 drops of the solution of NaCl in a clean, correctly marked test tube. Add 3 drops of a solution of ammonia (NH_3). Use a clean, dry stirring rod to stir the solution. Remove the stirring rod, and touch the adhering drop of solution to a small piece of red litmus paper. If the paper does not turn blue, add drops of NH_3 to the solution in the test tube, with stirring, until it does.

   CAUTION: Ammonia can cause chemical burns, in addition to ruining your clothing. If you spill any of the solution on you, wash the contaminated area thoroughly and report the incident to your laboratory instructor. You may require further treatment.

3. Repeat Step 2 with each of your solutions.
4. Add 5 drops of the solution of $Ba(NO_3)_2$ to each test tube. Shake gently to obtain homogeneity. Do not use your finger as a stopper. Examine each test tube carefully and look for precipitates. Record your observations, noting the color of each precipitate.

   CAUTION: Wash your hands thoroughly after using the solution containing barium, because it is poisonous.

5. The test in this step applies only to those test tubes that contain precipitates. Add 10 drops of the solution of HNO_3 to each of these test tubes. Test each solution with blue litmus paper by using a clean, dry stirring rod to transfer one drop of the solution onto the paper. If the paper does not turn pink, add drops of the solution of HNO_3 to the solution in the test tube, with stirring, until it does. Examine each test tube. Which precipitates have dissolved? Record your observations.
6. Discard the solutions in the test tubes.
7. Wash the test tubes, and rinse them with distilled water.

Testing with Silver Nitrate

1. Use fresh solutions for these tests.
2. Add 20 drops of the solution of NaCl to a clean, correctly marked test tube.
3. Repeat Step 2 with each of your solutions.
4. Add 5 drops of the solution of $AgNO_3$ to each test tube. Record your observations.
5. The test in this step applies only to those test tubes that contain precipitates. Add 10 drops of the solution of HNO_3 to each of these test tubes, and shake gently. Test a drop of each solution with blue litmus paper. If the paper does not turn pink, add drops of the solution of HNO_3 to the solution in the test tube, with stirring, until it does. Did the precipitates dissolve? Record your observations.
6. Discard the solutions in the test tubes. Wash the test tubes, and rinse them with distilled water.

Testing with Thymol Blue

1. Use fresh solutions for these tests.
2. Add 20 drops of the solution of NaCl to a clean, correctly marked test tube.
3. Repeat Step 2 with each of your solutions.
4. Add 5 drops of the solution of thymol blue to each test tube. Shake each test tube gently before recording your observations.
5. Discard the solutions in the test tubes. Wash the test tubes, and rinse them with distilled water.

Identifying the Unknown Compound

1. With one possible exception, the characteristic reactions of the unknown compound should be identical to those of one of the known samples. If your unknown is Na_2CO_3, its behavior in solution when tested with $AgNO_3$ may differ slightly from that of a known sample. Differences in the concentrations of these solutions can lead to reactions that result in slightly different colors. Keeping this possible difference in mind, you will be able to identify your unknown compound by matching its reactions with those of one of the known samples.
2. If ambiguities occur, repeat as many of the tests as you find necessary.

Date	________________	Student Name	________________
Course/Section	________________	Team Members	________________
Instructor	________________		________________

Identification of an Unknown Compound

Prelaboratory Assignment

1. Match the name of each compound with its formula.

sodium hydrogen phosphate	NaI
ammonia	HNO_3
sodium carbonate	NaCl
sodium sulfate	Na_2CO_3
nitric acid	Na_2HPO_4
sodium chloride	Na_2SO_4
sodium iodide	NH_3

2. List the signals for chemical reactions that you will find in this experiment.

3. What special safety precautions must be observed during this experiment?

Date ______________________ Student Name ______________________
Course/Section ______________________ Team Members ______________________
Instructor ______________________ ______________________

Identification of an Unknown Compound

Results

	HNO_3	$Ba(NO_3)_2$	$Ba(NO_3)_2$ + HNO_3	$AgNO_3$	$AgNO_3$ + HNO_3	Thymol Blue
NaCl						
NaI						
Na_2CO_3						
Na_2HPO_4						
Na_2SO_4						
Unknown No.______						

My unknown is__________________________.

Questions

1. When an unknown sample that may contain one of the five compounds from this experiment is treated with a solution of $AgNO_3$, a yellow precipitate forms.

   a. Using your record of observations, identify the unknown if possible. Note, however, that your record may show that more than one of the compounds is implicated. Explain your answer carefully.

   b. If more than one compound is implicated, how could you distinguish among them by using the tests in this experiment?

2. An unknown sample contains at least two of the five compounds from this experiment. Use the following observations to identify the components of the mixture. Explain your reasoning in a brief sentence.

   a. No gas is evolved when the solid is treated with HNO_3. The solution that results is colorless.

   b. A white precipitate forms after the addition of NH_3 and $Ba(NO_3)_2$ to the solution of the unknown. The precipitate does not dissolve after the addition of HNO_3.

   c. A white precipitate forms after the addition of $AgNO_3$ to a solution of the unknown. The precipitate does not dissolve after the addition of HNO_3.

   d. A yellow color results when thymol blue is added to a solution of the unknown.

1B. Separation of a Mixture by Paper Chromatography

Introduction

The separation, detection, and identification of the components of a mixture can be accomplished by several techniques. Each of these techniques depends on the differing chemical or physical properties of the components of the mixture. *Chromatography* (Ebbing/Gammon, Section 1.4) is one such technique. *Paper chromatography*, which is used here, is just one of several chromatographic methods available.

Purpose

In this experiment you will learn to separate, detect, and identify the components of a mixture by using paper chromatography. The components will be cobalt(II) chloride, nickel(II) chloride, copper(II) chloride, and possibly iron(III) chloride.

A Simple Explanation of Paper Chromatography

In paper chromatography, a drop of solution containing a mixture of substances is placed near one end of a rectangular piece of filter paper. The paper serves as the *stationary phase*. The end of the paper is immersed in a liquid to a point that is just below the spot where the drop was placed on the paper. The liquid is the *mobile phase*. Capillary action (the same phenomenon that causes water to travel up a bath towel when an edge of the towel is immersed) causes the liquid to flow up the filter paper. When the liquid reaches the spot, the components of the mixture will begin to migrate upward with the mobile phase. Each component will have a characteristic chemical affinity for the paper and a characteristic *chemical affinity* for the liquid. These affinities are competitive. The component's affinity for the paper tends to hold the component in one place, but its affinity for the liquid tends to make the component follow the liquid as it moves upward. A component with a strong affinity for the paper and a weak affinity for the liquid will move more slowly than a component with a weaker affinity for the paper and a stronger affinity for the liquid.

A substance's affinities for the stationary and mobile phases are entirely characteristic of that substance. Different substances have different competitive affinities. Because each component of a mixture has its own characteristic affinities, each component will travel up the paper at its own characteristic rate. If the paper is sufficiently large, all the components can be separated by the time the liquid has reached the top of the paper.

Each component will now appear as a separate spot. If the components are highly colored, the spots will be visible. You can convert weakly colored or colorless spots to highly colored spots by spraying them with substances that react with the components in the spots. The filter paper will now contain a vertical array of colored spots arranged according to their characteristic rates of ascent. The word *chromatography*, which is derived from two Greek words and literally means "written in color," was coined to describe this phenomenon. The distance traveled by a component of a spot with respect to the distance traveled by the pure liquid is a measure of that component's competitive affinities for the stationary and mobile phases. We define the component's R_F (retention factor) value in those terms:

$$R_F = \frac{\text{distance traveled by spot}}{\text{distance traveled by liquid}}$$

The R_F value of a substance is characteristic of that substance and should be a constant under invariant experimental conditions.

Concept of the Experiment

You will examine the paper chromatography of $CoCl_2$ [cobalt(II) chloride], $NiCl_2$ [nickel(II) chloride], and $CuCl_2$ [copper(II) chloride]. In addition, your laboratory instructor may elect to include $FeCl_3$ [iron(III) chloride] if your distilled water and reagents are not already so contaminated with this substance that the detection of a small quantity will be virtually impossible.

The mobile phase will be a mixture of an aqueous solution of HCl (hydrochloric acid) and either acetone or 2-butanone.

CAUTION: Because of the volatility and flammability of acetone and 2-butanone, no flames will be allowed in the laboratory.

At the completion of the experiment, you will spray the paper successively with solutions of NH_3 (ammonia) and Na_2S (sodium sulfide). The former will react with the HCl in the mobile phase to form NH_4Cl (ammonium chloride) according to the equation

$$NH_3 + HCl \rightarrow NH_4Cl$$

Sodium sulfide will react with a component of a spot to give a darkly colored spot containing one of the sulfides (Fe_2S_3, CoS, NiS, or CuS) and colorless sodium chloride (NaCl).

CAUTION: Hydrochloric acid and ammonia can cause chemical burns, in addition to ruining your clothing. If you spill one of these substances on you, wash the contaminated area thoroughly and report the incident to your laboratory instructor. You may require further treatment.

You will be able to determine the R_F value for each substance by observing its ascent (distance it travels up the paper) in the absence of the other substances. You will also subject a known mixture of all the substances to chromatography so that you can see that the same R_F values are obtained with a mixture. Finally, you will be able to identify the components in an unknown mixture on the basis of its chromatography and derived R_F values. This unknown will contain one or more of the substances whose behavior you have studied in this experiment.

Procedure

Getting Started

1. Obtain your unknown mixture.

2. Obtain directions from your laboratory instructor for discarding the solution that you will use as the mobile phase in this experiment.

3. Obtain a 5 × 9 inch piece of filter paper, a smaller scrap of the same paper, 5 capillary tubes, 4 test tubes, a piece of clear plastic film, a rubber band, and an 800-mL beaker. If your laboratory instructor elects to use $FeCl_3$, obtain 6 capillary tubes and 5 test tubes.

4. Add 35 mL of either acetone or 2-butanone to the 800-mL beaker, followed by 10 mL of the 6 *M* HCl solution. Swirl the beaker gently to mix the solution. Cover the beaker with the plastic film, and hold the film in place with the rubber band.

5. Place the large piece of filter paper on a clean paper towel. Use a pencil (not a pen) and a ruler to draw a line 2 cm from one of the longer edges.

6. Fold the paper in half so that the line you have drawn is bisected. In the same manner, fold the paper in half a second and then a third time. The line will have been divided into eight equal segments. Refold the paper so that it looks like an accordion, as shown in Figure 1B.1.

FIGURE 1B.1

A piece of chromatography paper that has been properly folded for use in this experiment.

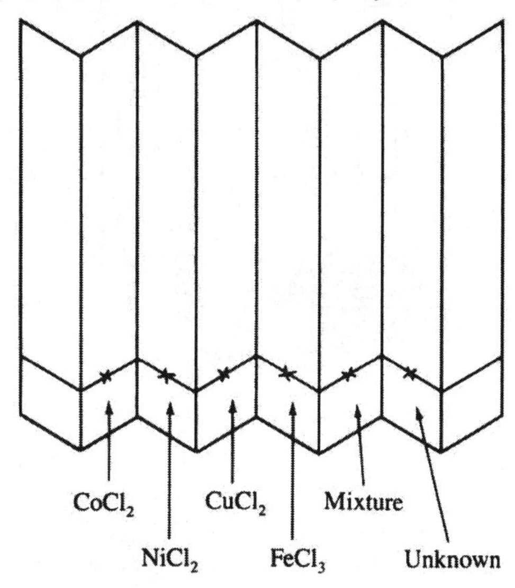

7. Mark the approximate center of each of the six inner segments of the line with the pencil. There will be no mark in the first segment on the left and none in the first segment on the right.

8. Write "$CoCl_2$" under the first mark on the left, "$NiCl_2$" under the next mark, "$CuCl_2$" under the third mark, "$FeCl_3$" under the fourth mark (if this substance is included in the experiment; leave it blank otherwise), "Mixture" under the fifth mark, and "Unknown" under the last mark.

9. Add some distilled water to one of the test tubes. Use one of the capillary tubes to practice spotting the scrap of filter paper. The maximum diameter of an acceptable spot is no larger than about 0.5 cm. When your procedure is satisfactory, empty the test tube and dry it. Dry the capillary tube by touching its wet tip to a piece of paper towel.

Doing the Experiment

1. Mark each of the test tubes for recognition with the name of one of the compounds to be tested. Place a small amount (about 1/4 inch) of the appropriate solution in the corresponding test tube.

2. Open the large piece of filter paper and place it once again on the paper towel. Use a clean capillary tube to spot the paper, placing the appropriate solution on the corresponding mark on the

   2-cm line.

3. Dry the paper by holding it *briefly* over a heat lamp or by waving it gently in the air. Refold the paper like an accordion.

4. Open the 800-mL beaker and gently place the folded paper inside. The 2-cm line should be above the surface of the liquid or close to it. Do not splash. Replace the plastic film. The result is shown in Figure 1B.2.

FIGURE 1B.2

The experimental arrangement, showing the covered beaker containing the folded chromatography paper (the stationary phase) and the solution (the mobile phase).

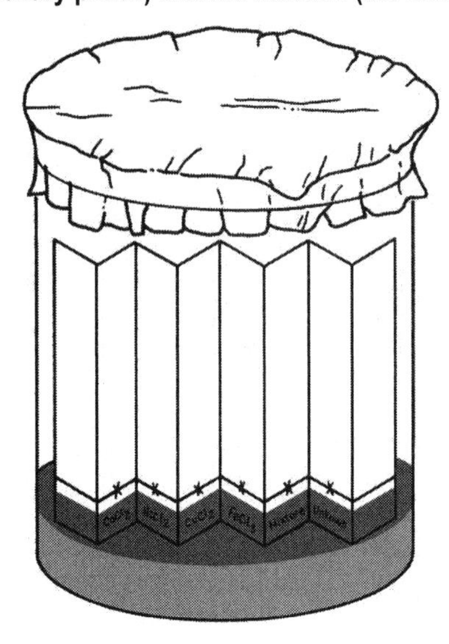

5. The beaker must be absolutely stationary throughout the experiment. To view the paper in its entirety, move to another position rather than turning the beaker.
6. Allow the liquid to ascend to within 2–3 cm of the top of the paper (about 30–40 minutes will be required). You should be able to observe the progress of some of the spots during the ascent. Record your observations, including the colors of the spots that you are able to see.
7. When the liquid has reached the desired height, remove the paper from the beaker. Place the wet paper on the paper towel, and mark with a pencil the position to which the pure liquid has ascended.
8. Partially dry the paper, using the drying methods described in Step 3.
9. The operation described in Step 10 must be done in a hood.

   CAUTION: Sodium sulfide and its reaction product with water are toxic substances that should not be tasted or inhaled. An efficient hood must be used during this step.

10. Wearing rubber or plastic gloves, spray the paper with a solution of NH_3. The paper should be moist but not wet. Next spray the paper with a solution of Na_2S. A dark spot should now be visible for each substance. Dry the paper.
11. Circle each spot with a pencil.
12. Measure the vertical distance that the approximate center of each of these spots has traveled from the original mark on the 2-cm line. Your measurements should be to the closest tenth of a centimeter. Record your results.
13. Measure and record the vertical distance from the 2-cm line that the pure liquid has ascended.

    Calculate the R_F value for each spot.
14. Use the R_F values to deduce the identity of each component in the unknown mixture.
15. Attach the dry chromatography paper to your report.

Date ____________________ Student Name ____________________
Course/Section ____________________ Team Members ____________________
Instructor ____________________ ____________________

Separation of a Mixture by Paper Chromatography

Prelaboratory Assignment

1. Match the name of each compound with its formula.

copper(II) chloride	$NiCl_2$
sodium sulfide	HCl
iron(III) chloride	NH_3
nickel(II) chloride	$CuCl_2$
cobalt(II) chloride	Na_2S
ammonia	$CoCl_2$
hydrochloric acid	$FeCl_3$

2. A piece of filter paper is spotted with a solution containing a mixture of two components, A and B. The chemical affinity of A for the stationary phase is less than that of B, and the chemical affinity of A for the mobile phase is greater than that of B. Which substance will have traveled farther at the completion of the chromatography experiment? Which substance will have the larger R_F value? Explain each answer.

3. What special safety precautions must be taken during this experiment?

Date		Student Name	
Course/Section		Team Members	
Instructor			

Separation of a Mixture by Paper Chromatography

Results

	Observations	Distances	R_F Values
$CoCl_2$			
$NiCl_2$			
$CuCl_2$			
$FeCl_3$			
Mixture			
Unknown No. ______			

Distance solvent traveled: __________cm

Components of the unknown mixture: __________

Questions

1. Why is a pencil, rather than a pen, used to draw the lines on your piece of filter paper? (See, for example, Ebbing/Gammon, Figure 1.17.)

2. Iron(III) chloride is a common impurity in this experiment because the distilled water in the solution of HCl may have passed through iron pipes. Examine your piece of filter paper carefully. If this impurity is present, why is it a horizontal streak rather than a spot? If it is not present, why would its presence as an impurity result in a horizontal streak of color rather than a spot?

3. Write balanced equations for the reactions of $FeCl_3$, $CoCl_2$, $NiCl_2$, and $CuCl_2$ with Na_2S.

1C. Some Measurements of Mass and Volume

Introduction

Many experiments require some type of measurement, and often these are simple measurements of mass and volume. The validity of an experiment is likely to depend on the reliability of these measurements. A measurement's reliability is usually considered in terms of its *accuracy* and *precision* (Ebbing/Gammon, Section 1.5). The relationship among accuracy, precision, and error is discussed in Appendix A to this manual. The relationship between precision and significant (or meaningful) figures can also be found there.

After a quantity has been measured in an experiment, it may be necessary to use that measurement in a subsequent calculation. If a hand calculator is used for the arithmetic, eight or more digits may appear in the answer. Are all of these digits meaningful? Two simple rules for determining the number of significant figures that should appear in an answer are given in Appendix A.

Purpose

This experiment will give you an opportunity to measure the mass of an object with a balance and to calculate the volume of a liquid that is delivered from a transfer pipet. You will also determine the precision of your measurements. The experiment also includes an optional calculation of standard deviations using a computer and the Internet.

Balances

Figure 1.C1 and Figure 1C.2 show two commonly used types of balances, but other types of balances are in use. Some balances allow a precision of ±0.001 g, whereas others offer considerably less. Your laboratory instructor will provide details about the operation of the balances in your laboratory and the precision that you can expect. You should be able to achieve the maximum precision offered by your balance almost immediately.

FIGURE 1C.1

A top-loading, digital balance.

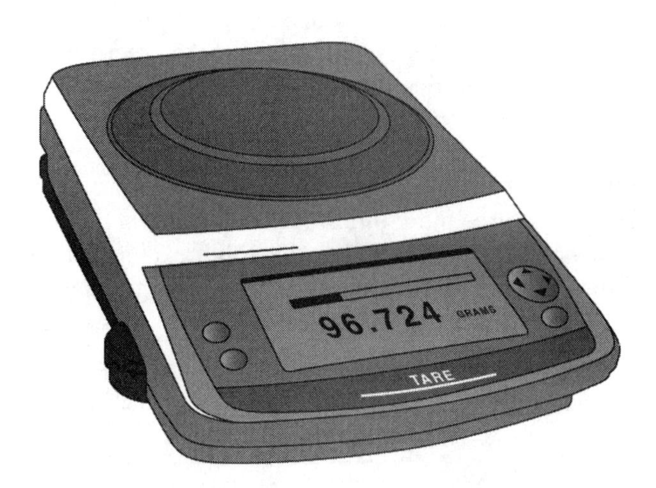

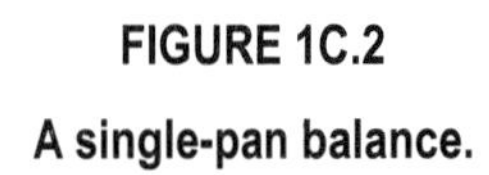
FIGURE 1C.2

A single-pan balance.

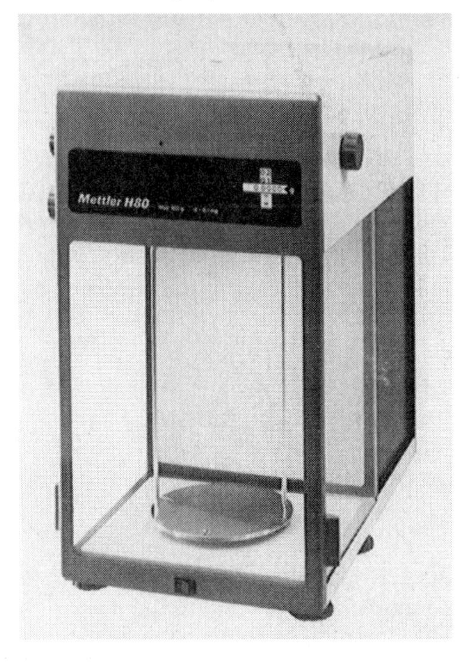

Transfer Pipets

A transfer pipet is calibrated to deliver a specified volume of a liquid. Correct use of this piece of glassware requires a good deal of manipulatory skill. As with any other skill, practice is mandatory. The precision that you can achieve with a pipet will depend on how much time you devote to practice.

The correct use of a pipet is discussed in the Introduction to this manual; it may also be demonstrated by your laboratory instructor. These instructions will make it clear that you will be filling the pipet with a liquid taken from one container and allowing the liquid to drain from the pipet into another container.

> **CAUTION: Never use your mouth to draw a liquid into the pipet, even if the liquid is water. Use a rubber suction bulb or some other suction device.**

Concept of the Experiment

You will be able to practice using a balance and a transfer pipet in order to gain confidence. Next you will measure the mass of a flask four times. You will examine the precision of your measurements when you determine the correct number of significant figures in the mean mass of the flask, using the method discussed in Appendix A.

You will add water to the flask from a filled 10-mL pipet and then measure the mass of the flask and water. You will repeat this process three more times. After calculating the mass of the water that was delivered each time from the pipet, you will calculate the volume of each addition from the mass and density of water. You will then determine the correct number of significant figures in the mean volume. This number will allow you to appreciate the precision that you have achieved with the pipet.

Procedure

Getting Started

1. Obtain about 100 mL of distilled water in a beaker. Allow the beaker and water to sit on the laboratory bench while you are learning to use the balance and the pipet. The water should come to the temperature of the laboratory during that time.
2. Obtain also a 10-mL pipet, a thermometer, and a 50-mL Erlenmeyer flask with a rubber stopper.
3. Plan on using the same balance and pipet throughout the experiment.

Using Your Balance

1. Ask your laboratory instructor for the maximum precision offered by your balance. That precision is ±___g.
2. Obtain instructions for using your balance, and practice using it by measuring the mass of an object (such as a coin) several times.
3. Place the rubber stopper in the Erlenmeyer flask. Bring your balance to the zero or null position.

   Measure and record the combined masses of the flask and stopper.
4. Use tissue paper to remove the stoppered flask from the pan of the balance. (The tissue paper is used because some balances are sensitive enough to detect the oils from your fingerprints.)
5. Bring your balance to the zero or null position again. Measure and record the mass of the stoppered flask once more.
6. Repeat Steps 4 and 5 until you have measured the mass four times.
7. Calculate the mean mass.
8. The differences between the measured masses and the mean should be very small. Ask your laboratory instructor whether your results are satisfactory before you proceed.

Using the Pipet

1. Practice with your pipet using distilled water (do not use the water you have set aside) until you are comfortable with the technique.
2. Using the thermometer, note the temperature of the laboratory and of the distilled water that you have set aside. When the temperatures are identical or very nearly identical, you are ready to begin. Record the temperature to the nearest degree.
3. Measure and record the mass of the stoppered flask again. Use tissue paper as you did before.
4. Remove the flask from the balance, using tissue paper. Remove the stopper, using tissue paper.

   Pipet 10 mL of the room-temperature water into the flask without touching the flask with your fingers. Using tissue paper, replace the stopper to prevent evaporation.
5. Bring your balance to the zero or null position. Measure and record the combined mass of the water and the stoppered flask.
6. Remove the flask from the balance. Using the same precautions, pipet another 10-mL sample into the flask. Do not pour out the first sample. The volume of water in the flask should now be 20 mL. Replace the stopper and repeat Step 5.
7. Repeat until four samples of water have been delivered to the flask and the final volume is 40 mL.

8. Calculate the mass of water that was delivered each time from your pipet. These masses should be approximately identical.

Table 1C.1

Density (g/ mL) of Water at Various Temperatures (°C)

Temp.	Density	Temp.	Density	Temp.	Density
17	0.9988	22	0.9978	27	0.9965
18	0.9986	23	0.9976	28	0.9962
19	0.9984	24	0.9973	29	0.9959
20	0.9982	25	0.9971	30	0.9956
21	0.9980	26	0.9968	31	0.9953

9. Calculate the volume of each sample from the mass and density of water. Use the density in Table 1C.1 that corresponds to your recorded temperature. Due regard for significant figures (Appendix A) should be observed in your calculations. Do not attempt yet to limit the number of significant figures on the basis of the precision of your measurements.

Date	______________________	Student Name	______________________
Course/Section	______________________	Team Members	______________________
Instructor	______________________		______________________

Some Measurements of Mass and Volume

Prelaboratory Assignment

1. Compare and contrast accuracy and precision.

2. a. Compare and contrast mistake and error.

   b. Compare and contrast systematic and random errors.

3. a. What is the rule concerning significant figures when measured quantities are added or subtracted (Appendix A)?

b. What is the rule concerning significant figures when measured quantities are multiplied or divided (Appendix A)?

4. Suppose that a series of measurements have shown that the mean mass of an object is 122.4 g with a standard deviation of 1.3 g. How many significant figures are justified in the mean and how would you report the most probable mass of the object? Explain.

5. What safety rule concerning the pipet must be observed during this experiment?

Date ______________________ Student Name ______________________
Course/Section ______________________ Team Members ______________________
Instructor ______________________ ______________________

Some Measurements of Mass and Volume

Results

1. *Using the analytical balance*

Mass of the stoppered flask (g) __________ __________ __________ __________

Mean mass (g) __________

Calculation:

2. *Using the pipet*

Temperature (°C) __________

Addition No.	**1**	**2**	**3**	**4**
Mass *after* addition (g)	__________	__________	__________	__________
Mass *before* addition (g)	__________	__________	__________	__________
Mass of added water (g)	__________	__________	__________	__________
			Density of water (g/mL)	__________
Volume of water delivered each time (mL)	__________	__________	__________	__________
			Mean volume (mL)	__________

Calculations:

Questions

1. a. Calculate the standard deviation in the measured mass of the empty stoppered flask (see Appendix A).

   b. Determine how many significant figures in the mean measured mass of the flask are allowed by the precision of the mass measurement (see Appendix A). Give the mean mass with the correct number of significant figures.

   c. How does this result compare with the claimed precision of your balance?

   d. What, if anything, can you claim about the accuracy of the mean mass?

Student name: _______________ Course name: _______________ Date: _____________

2. a. Using the mean volume of water and the standard deviation, determine the number of significant figures that is allowed by the precision inherent in your pipet technique. Give the mean volume with the correct number of significant figures.

b. Some claim that a precision of ±0.01 mL can be achieved with a 10-mL pipet. However, students who are learning the technique of using a pipet may not achieve this precision. How do your results compare with the claim? If you have not been able to obtain this precision, try to pinpoint the deficiencies in your pipet technique.

3. With due regard for significant figures, determine what mass of water would be delivered from your pipet at 39°C. The density of water at that temperature is 0.9918 g/mL.

2. Isotopes and Mass Spectrometry

Introduction

We owe our knowledge of the existence of isotopes and the isotopic composition of the elements to the invention and use of *mass spectrometers* (Ebbing/Gammon, Section 2.4). In addition, these ingenious devices can be used to determine the masses of isotopes in elements as well as in compounds.

Purpose

Although you will not need a mass spectrometer, you will use the mass spectrum of neon, as well as the known abundances of this element's naturally occurring isotopes, to devise a method for obtaining isotopic abundances from mass spectra. You will then apply your method to the mass spectra of mercury, hydrogen chloride, and hydrogen bromide to determine the isotopic composition and atomic weights of mercury, chlorine, and bromine. Your instructor may also ask you to explain the mass spectrum of elemental bromine.

Isotopes

An element is distinguished from any other element by its atomic number. All the atoms of a particular element have the same atomic number. An atom is also characterized by its mass number. Although the atoms of a particular element have only one atomic number, they may have two or more different mass numbers. *Isotopes* are atoms that have the same atomic number but different mass numbers.

The atoms of most of the elements consist of two or more isotopes. For example, neon consists of three naturally occurring isotopes: ^{20}Ne (10 protons and 10 neutrons), ^{21}Ne (10 protons and 11 neutrons), and ^{22}Ne (10 protons and 12 neutrons). The abundances of these isotopes are 90.51%, 0.27%, and 9.22%, or, in terms of fractional abundance, 0.9051, 0.0027 and 0.0922, respectively. Thus 9051 of every 10,000 neon atoms are ^{20}Ne, 27 are ^{21}Ne, and 922 are ^{22}Ne.\

The *atomic weight* of an element is the weighted average of the exact masses (in atomic mass units) of the naturally occurring mixture of isotopes. Each exact isotopic mass must be multiplied by the fractional abundance of the isotope, and each result must be added to the others. For an element Y that consists of isotopes ^{a}Y, ^{b}Y, and so on, the atomic weight of Y is given by

Atomic weight of Y = (exact mass ^{a}Y) × (fractional abundance ^{a}Y) + (exact mass ^{b}Y) × (fractional abundance ^{b}Y) + ...

Using this type of calculation, it can be shown that the atomic weight of neon is 20.18 amu.

Mass Spectra of Atoms

The way a mass spectrometer works is described in your textbook (Ebbing/Gammon, Section 2.4). An atom (X) in the gas phase collides with a high-energy electron and loses an electron:

$$X + e^- \rightarrow X^+ + 2e^-$$

The positive ion (X^+) moves through the mass spectrometer under the combined influences of an electric field and a magnetic field until it arrives at a collection point. A special detecting device counts the ions as they arrive at this point.

The mass of the ion, rather than its chemical nature, determines its exact path. As a result, all of the ions with the same mass are counted together by the detector. Ions with a different mass arrive at a different place at the detector and so are counted separately. The total number of counts for each ion is recorded on a graph called a *mass spectrum*. Two or more of these graphs are called *mass spectra.*

Consider the mass spectrum of neon shown in Figure 2.1. It consists of an intense signal at mass number 20, a much smaller signal at mass number 21, and a moderately intense signal at mass number 22. You will recall that these are the mass numbers of the three isotopes of neon. *The heights of these signals are proportional to the number of counts at each mass number and, in turn, reflect the fractional abundances of the isotopes.* Make sure you understand this concept by comparing the known fractional abundances of neon's isotopes with the heights of the appropriate signals in the mass spectrum.

Figure 2.1

The mass spectrum of neon.

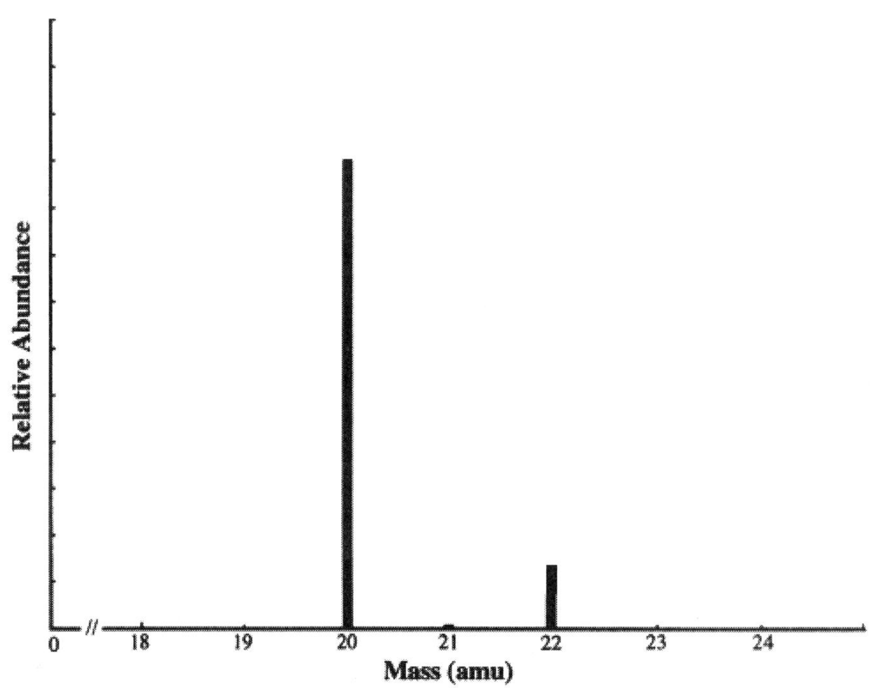

Mass Spectra of Molecules: Fragmentation

When a high-energy electron collides with a molecule, the result is usually a little more complicated than when it collides with an atom. Much of the colliding electron's energy is given to the molecule. If this energy is sufficient, a chemical bond may break and the molecule flies apart. Each fragment can then acquire a positive charge when it is struck by another electron. Thus the mass spectrum of a molecule contains signals that are due to fragments of molecules in the mass spectrum, as well as those that are due to the unfragmented parent molecules.

As an example, consider the mass spectrum of hydrogen iodide (HI) shown in Figure 2.2. The signal at mass number 128 is due to HI^+ ions because iodine consists solely of ^{127}I atoms and the mass number of hydrogen is 1 (127 + 1 = 128). Note that ^{1}H is the only isotope of hydrogen that affects the mass spectrum because the isotopic abundance of the other naturally occurring isotope, ^{2}H (deuterium), is only 0.015%. Finally, the signal at mass number 127 in Figure 2.2 comes from I^+ ions that result from fragmentation of the parent molecule. Note that we do not try to explain the relative intensities of signals due to a parent ion such as HI^+ and its fragmentation product, I^+. *We reserve our interpretation of the relative intensities of two or more signals to those arising from isotopes.*

Figure 2.2

The mass spectrum of hydrogen iodide (HI).

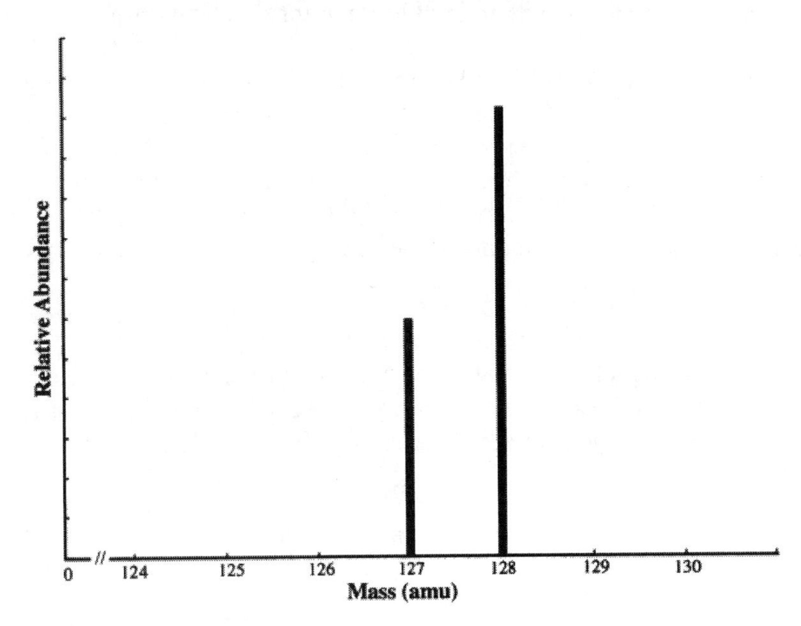

Concept of the Experiment

You will devise a method for determining the fractional abundances of neon's isotopes from the mass spectrum in Figure 2.1. Remember that the height of a signal is proportional to the fractional abundance. Because you know the correct fractional abundances before you begin, you will be able to judge the worth of any method you devise.

You will then be able to use your method to obtain the fractional abundances of the isotopes of mercury from the mass spectrum of this element. Mercury has naturally occurring isotopes with mass numbers of 196, 198, 199, 200, 201, 202, and 204.

Next you will obtain the isotopic abundances of the isotopes of chlorine from the mass spectrum of hydrogen chloride. The isotopes of chlorine have mass numbers of 35 and 37, so a sample of HCl contains both $H^{35}Cl$ and $H^{37}Cl$ molecules. The relative abundances of these molecules will be identical to the fractional abundances of the ^{35}Cl and ^{37}Cl isotopes. Thus the relative intensities of the signals arising from $H^{35}Cl^+$and $H^{37}Cl^+$ions will reflect the fractional abundances of the two isotopes of chlorine. In addition, you will see signals from $^{35}Cl^+$and $^{37}Cl^+$ions coming from the fragmentation of $H^{35}Cl^+$and $H^{37}Cl^+$.

In a similar fashion, you will obtain the isotopic abundances of the isotopes of bromine from the mass spectrum of hydrogen bromide. However, you will be required to deduce the mass numbers of bromine's isotopes from the mass spectrum.

You must provide an explanation for the origin of each signal in the spectra before you can calculate the fractional abundances of chlorine and bromine.

You will be required to calculate the atomic weights of mercury, chlorine, and bromine from your results. For purposes of this experiment, you will use mass numbers instead of exact masses.

Finally, if your instructor elects, you will be given an opportunity to explain the mass spectrum of elemental bromine. This substance consists of diatomic (two-atom) molecules, Br_2. All of the halogens (fluorine, chlorine, bromine, iodine, and astatine) are diatomic. As you will see if you do this part of the experiment, interpreting the mass spectrum of a diatomic element with more than one isotope requires considerably more insight than interpreting that of a monatomic (one-atom) element such as neon.

Procedure

Getting Started

1. Your laboratory instructor may ask you to work with a partner.
2. Obtain the measuring device that you require.

Doing the Experiment

1. Give the origin of each signal in the mass spectra of mercury and hydrogen chloride. These spectra can be found in Figures 2.3 and 2.4. Make the required measurements (including units) and record them. Use the same degree of precision with each measurement.

Figure 2.3

The mass spectrum of mercury.

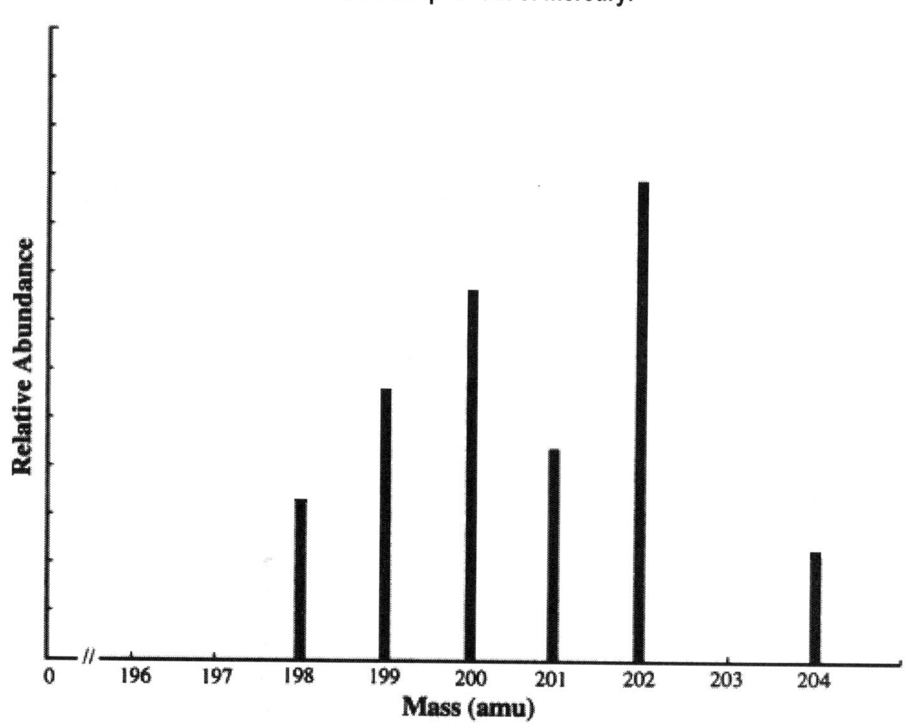

Figure 2.4

The mass spectrum of hydrogen chloride (HCl).

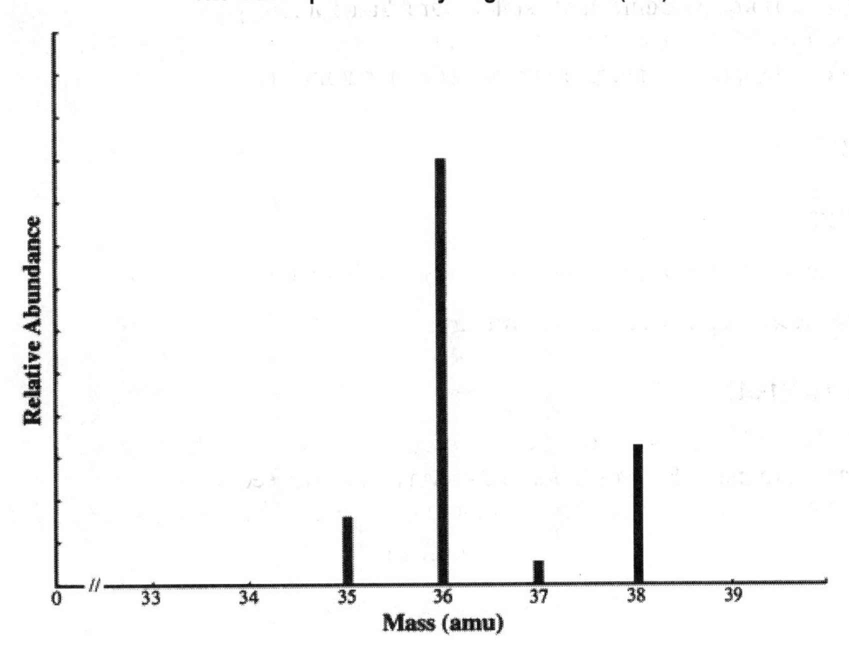

2. Consider the mass spectrum of hydrogen bromide (Figure 2.5). What isotopes of bromine would lead to this spectrum? Record your answers. Make and record the required measurements (including units).

3. If required by your laboratory instructor, consider the mass spectrum of elemental bromine (Br_2). This spectrum can be found in Figure 2.6. Make and record the required measurements (including units).

Figure 2.5

The mass spectrum of hydrogen bromide (HBr).

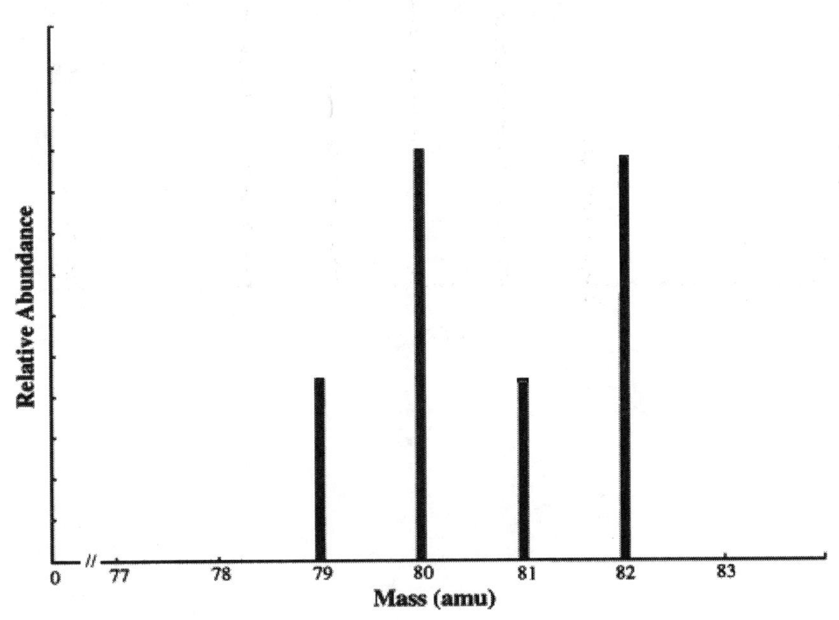

Figure 2.6

The mass spectrum of elemental bromine (Br_2).

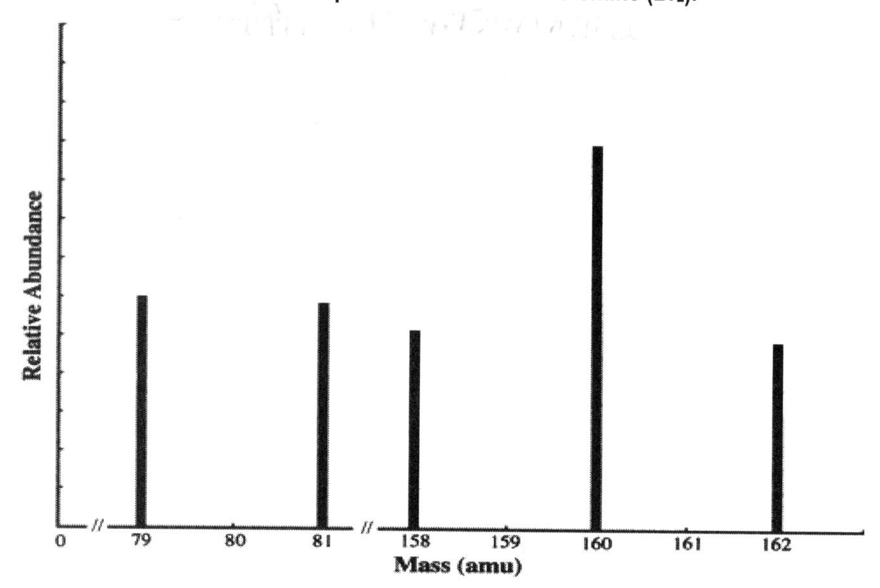

Date		Student Name	
Course/Section		Team Members	
Instructor			

Isotopes and Mass Spectrometry

Prelaboratory Assignment

1. Suppose the mass spectrum of a hypothetical monatomic element X contains a signal at mass number 13 and another of identical height at mass number 15.

   a. Sketch the mass spectrum. Make sure each axis is properly labeled.

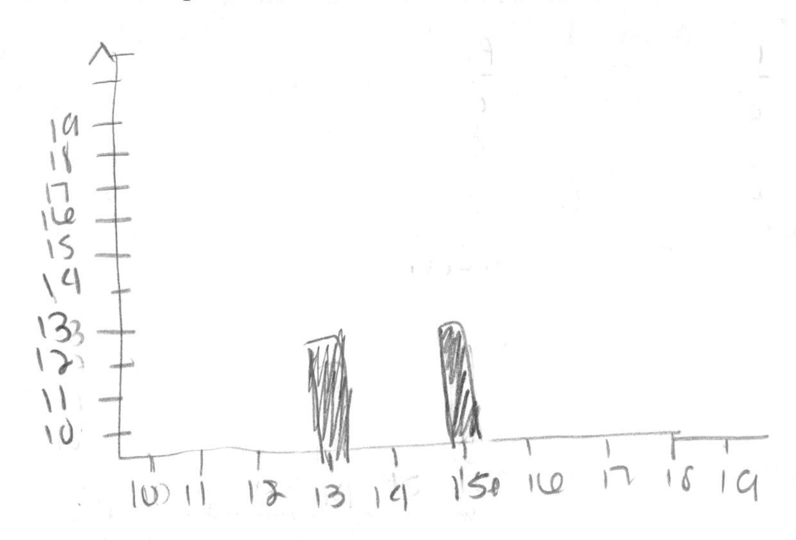

   b. How many isotopes are present? Why?

   There are 2 isotopes presented because they are both different

c. What are the fractional abundances of the isotopes? Why?

The fractional abundance is 0.5 cm according to my calculations.

2. a. Devise a *general* method for determining the fractional abundances of two or more isotopes from a mass spectrum. Your method must include some means of measuring the signals. You must state the type of measuring device that you will need to apply your method. You must also indicate how you will use the measurements to obtain the fractional abundances.

individual length / total = fraction / abundance

I will use a ruler + measure in cm.
I will use the individual measurement.

b. Test your method by determining the fractional abundances of neon's isotopes from the mass spectrum in Figure 2.1. When you compare your results to known abundances (shown on page 43), remember to take the precision of the measurements into account.

Mass Number	Measurement with Units
20	10
21	0
22	2

Calculation:

20 Ne 10   22 Ne 10

Ne-20 = 0.841

FA = individual length / total   10 cm / n = 0.410 cm

Student name: ________________ Course/Section: ________________ Date: ________________

Comments:

3. a. What does "fragmentation" in a mass spectrometer mean?

In a mass spectrometer "fragmentation" means a high energy electron collided with a molecule and if the energy is sufficient the chemical bond will then break, next the molecule will separate.

b. How will fragmentation affect the mass spectrum of hydrogen chloride? Explain in one or two sentences..

There are different hydrogen and chlorine therefore the mass of each fragment will be different from the parent molecule. chlorine mass is smaller than the compound of HCl.

Date ______ Student Name ______
Course/Section ______ Team Members ______
Instructor ______ ______

Isotopes and Mass Spectrometry

Results

1. *Mercury peaks*

Mass Number	Measurements with Units
198	2.1 cm
199	3.6 cm
200	4.9 cm
201	2.8 cm
202	63 cm
204	1.5 cm

2. *Hydrogen chloride, HCl, peaks*

Mass Number	Formula	Measurements with Units
36	HCl-35	5.6 cm
35	Cl-35	1 cm
38	HCl-37	2.0 cm
37	Cl-37	0.3 cm

3. *Hydrogen bromide, HBr, peaks*

Mass Number	Formula	Measurements with Units
80	HBr - 79	5.4 cm
79	Br - 79	2.4 cm
82	HBr - 81	5.4 cm
81	Br - 81	2.4 cm

How many isotopes of bromine are indicated by the number of peaks in the mass spectrum? Explain how you came to this conclusion.

4. *Bromine, Br_2, peaks (optional)*

Mass Number	Formula	Measurements with Units

Student name: ________________ Course/Section: ________________ Date: ______________

Questions

1. Calculate the fractional abundances of the isotopes of mercury, chlorine, and bromine.

mercury

m - 198 - 0.4/21.2 = 0.10

m - 199 - 0.17/21.2 = 0.7

m - 200 - 4.9/21.2 = 0.23

m - 201 - 2.8/21.2 = 0.13 cm

m - 202 - 6.3/21.2 = 0.30 cm

m - 204 - 1.5/21.2 = 0.07 cm

2.1 3.5 4.9 + 2.8 + 6.3 + 1.5 cm

chlorine

5.6 + 1.0 = 6.6 cm

1.35 + 1.0 = 1.45

2.0 + 0.3 = 2.3 cm

6.6/8.9 = 0.74 cm

2.3/8.9 = 0.26 cm

Bromine Br - 79 - 7.8 Br - 81 - 7.8

5.4 + 2.4 = 7.8

7.8 + 7.8 = 15.6

2.4 ÷ 7.8 = 0.31

7.8 / 15.6 = 0.5 cm

7.8 / 15.6 = 0.5 cm

2. Using the method outlined on page 43, calculate the atomic weights of mercury, chlorine, and bromine from your data. Use mass numbers rather than exact masses. Compare your results to the actual atomic weights of these elements, and comment on any discrepancies.

mercury →
Hg - 198 x 0.10 = 19.8
Hg - 199 x 0.17 = 33.83 cm
Hg - 200 x 0.23 = 46.00 cm
Hg - 201 x 0.13 = 26.13 cm
Hg - 202 x 0.30 = 60.60 cm
Hg - 204 x 0.07 = 14.28 cm

19.8 + 33.83 + 46.00 + 26.13 + 60.60 + 14.28 = 200.64
atomic weight 200.64

chlorine →
Cl 35 x 0.74 = 25.9

Cl - 37 x 0.37 = 9.62

25.9 + 9.62 = 35.5
atomic weight = 35.5

Bromine →
Br - 79 x 0.5 = 39.5
Br - 81 x 0.5 = 40.5

39.5 + 40.5 = 80

Student name: _______________ Course/Section: _______________ Date: _______________

3. (Optional) Why does the mass spectrum of Br_2 contain three signals whose heights are almost in the ratio of 1:2:1? What combinations of isotopes are responsible for these signals? It may help to suppose that the fractional abundances of the isotopes are exactly equal. Then think about the probability of combining the various isotopes of bromine atoms into diatomic molecules. Finally, why does the spectrum contain two other signals of roughly equal height? What are the origins of these signals?

3A. The Empirical Formula of an Oxide

Introduction

Every chemical compound has a chemical formula that can be determined by a combination of experiments and calculations. For most ionic compounds, such as magnesium oxide, the *empirical formula* (the simplest formula) is the formula that is used for the compound. An empirical formula is the formula with the smallest integer (whole-number) subscripts (Ebbing/Gammon, Section 3.5).

Purpose

You will be able to determine the empirical formula for magnesium oxide from the results that you will obtain by burning magnesium in air.

What Happens When an Element Is Burned in Air?

Molecular oxygen, alone or in air, is a very reactive substance when it is heated. Many elements will react with it. When an element reacts and combines chemically with molecular oxygen, an *oxide* (a compound of the element with oxygen) is usually formed.

Molecular nitrogen, the chief component of air, is a rather unreactive substance, even at a high temperature. Only the more active metals will react and combine chemically with molecular nitrogen during heating. When nitrogen does react with an active metal, a *nitride* (a compound of the element with nitrogen) is formed.

Although the amount of molecular nitrogen in the air is approximately four times the amount of molecular oxygen, more oxide than nitride is formed when an active metal is burned in air. The reason is the superior reactivity of molecular oxygen.

Concept of the Experiment

In this experiment, you will burn a known amount of magnesium in air producing magnesium oxide and smaller amounts of magnesium nitride (Mg_3N_2). Water will convert the nitride to magnesium hydroxide [$Mg(OH)_2$] with the liberation of ammonia (NH_3). Heat will cause conversion of the hydroxide to the oxide with the loss of gaseous water.

Because the product will consist solely of the oxide after this treatment, you can determine the mass of oxygen that is present in the oxide from its mass and the original mass of the magnesium. The laws of conservation of mass, as well as the concept of a mole, will lead you to the method by which you can determine the empirical formula of this oxide (Ebbing/Gammon, Sections 1.3 and 3.2).

Procedure

Getting Started

1. Your laboratory instructor may wish to provide special safety precautions concerning the bulk supply of magnesium ribbon.
2. Ask your laboratory instructor about discarding the magnesium oxide that you will prepare in this experiment.
3. Obtain a crucible and lid. Wash, rinse, and dry them.

4. Obtain about 0.2 g of magnesium ribbon. If it is not bright, clean the surface with sandpaper.

5. Place the covered crucible in a clay triangle on an iron ring that is attached to a ring stand. Adjust the height of the ring so that the bottom of the crucible will be in the hottest part of a properly adjusted laboratory burner. (The Introduction to this manual discusses burners and their use.) The correct arrangement of the equipment, crucible, and burner is shown in Figure 3A.1.

FIGURE 3A.1

The correct arrangement of the ring stand, the clay triangle, the crucible with its lid, and the burner.

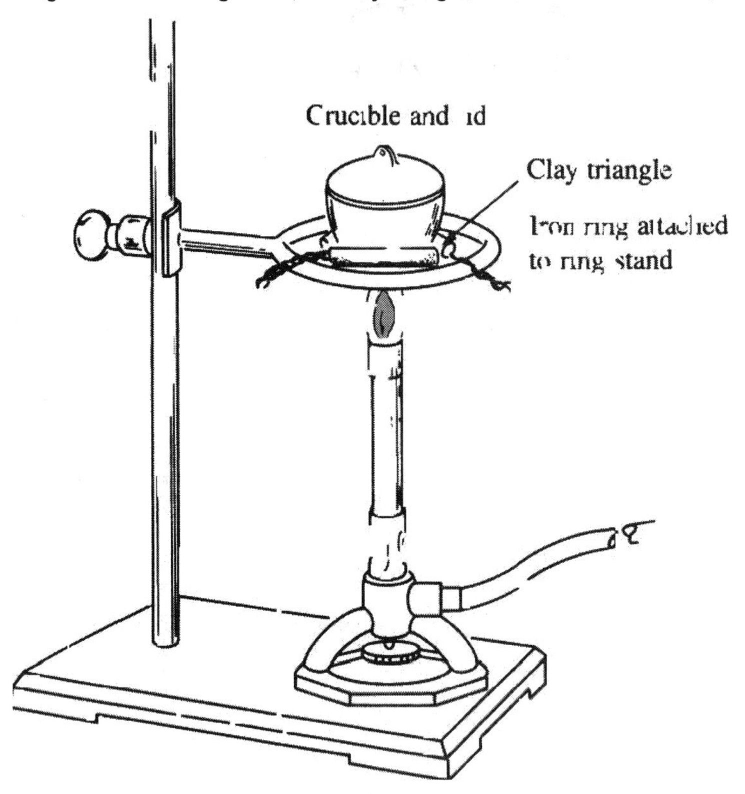

Doing the Experiment

1. Heat the covered crucible for about 3 min. The bottom of the crucible should attain a red-hot glow during this time. Move the burner and allow the crucible to cool (5–10 min). When the crucible is cool, you should feel no heat when you place one of your fingers about 1/2 inch from the bottom of the crucible.

CAUTION: Avoid burning your fingers. Do not touch the crucible or the iron ring at any time during this experiment.

FIGURE 3A.2

The proper method for carrying the crucible, with crucible tongs over wire gauze. A watch glass may be used instead of the wire gauze.

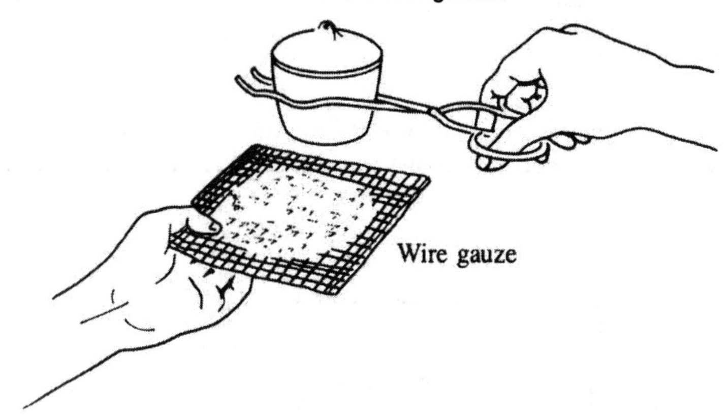

2. When the covered crucible is cool, transfer it to the pan of a balance using crucible tongs. When moving the crucible to the balance, hold a wire gauze under the crucible (Figure 3A.2), Do not put the wire gauze on the pan. If you must wait to use the balance, do not place the crucible directly on the bench. Put it on the wire gauze or leave it in the clay triangle.
3. Obtain and record the mass of the covered crucible.
4. Repeat Steps 1, 2, and 3 until two consecutive masses differ by no more than ±0.001 g or any other precision that is stipulated by your laboratory instructor. Record the mean, or average, value of these masses. You will use this result in subsequent calculations.
5. Fold the magnesium ribbon into a loose ball that will fit completely inside the crucible. Do not fold the ribbon too tightly. The best results will be obtained when as much of the surface of the ribbon as possible is exposed to the air in the crucible.
6. Cover the crucible, obtain the mass, and record it once again.
7. Return the crucible to the clay triangle, using crucible tongs and the wire gauze. The lid should be completely in place. Brush the bottom of the crucible with the flame for about 2 or 3 min. Next place the burner on the ring stand, and heat the crucible in the hottest part of the flame for another 3 min.
8. Use crucible tongs to lift the lid carefully by a slight amount to allow more air to enter the crucible. Do not open the lid too far because doing so will allow the metal to enflame (catch on fire). The metal should glow brightly without flames. Flames from the metal must be avoided because they will carry part of the solid oxide out of the crucible and into the air as smoke.
9. Continue Step 8 until no metal is evident and no glow occurs when the lid is lifted.
10. Allow the covered crucible and its contents to cool. The contents should be white or slightly gray.
11. Remove the lid and place it on the wire gauze. Add a few drops of distilled water from a medicine dropper directly on the contents. You may notice the smell of ammonia at this point.
12. Place the lid onto the crucible so that the lid is slightly ajar. Heat the crucible by brushing it with the flame until the contents are dry. Next heat the crucible in the hottest part of the flame for 8–10 min to convert the magnesium hydroxide to the magnesium oxide.
13. Allow the covered crucible and its contents to cool to the same point as in Step 1.
14. Obtain the mass of the covered crucible and magnesium oxide. Record the mass.
15. Heat the covered crucible strongly again for about 3 min. Obtain and record the mass after the

crucible has cooled.

16. You will have obtained a “constant” mass if the two measurements agree to the precision specified in Step 4. If they do not, repeat Step 15 until two successive measurements differ by no more than this amount.

17. Clean the crucible and lid carefully.

CAUTION: Before leaving the laboratory, make sure that your gas outlet and those of your neighbors are closed.

Date ______ Student Name ______
Course/Section ______ Team Members ______
Instructor ______ ______

The Empirical Formula of an Oxide

Prelaboratory Assignment

1. a. What is the law of definite proportions, and how does it apply to this experiment?

   b. What is a mole and what is a molar mass? What are the molar masses of magnesium and atomic oxygen?

   c. Why is magnesium the limiting reactant (Ebbing/Gammon, Section 3.8) in this experiment?

2. What safety precautions are cited in this experiment?

3. When 0.422 g of phosphorus is burned, 0.967 g of a white oxide is obtained.

   a. Determine the empirical formula of the oxide.

   b. Write a balanced equation for the reaction of phosphorus and molecular oxygen on the basis of this empirical formula.

Date ____________________ Student Name ____________________
Course/Section ____________________ Team Members ____________________
Instructor ____________________ ____________________

The Empirical Formula of an Oxide

Results

Mass of empty crucible and lid (g): ____________ ____________

____________ ____________

____________ ____________

Average mass of empty crucible and lid (g): ____________

Mass of crucible, lid, and Mg (g): ____________

Average mass of empty crucible and lid (g): ____________

Mass of Mg (g): ____________

Mass of crucible, lid, and magnesium oxide (g): ____________ ____________

____________ ____________

____________ ____________

Average mass (g) of crucible, lid, and magnesium oxide: ____________

Average mass of empty crucible and lid (g): ____________

Mass of magnesium oxide (g): ____________

Questions

1. Determine the empirical formula of the oxide of magnesium. (Think carefully about the precision of each mass that you measured before you do this calculation. How many significant figures are justified for each mass?)

2. Write correctly balanced chemical equations for the following reactions:

   a. Magnesium with molecular oxygen

   b. Magnesium with molecular nitrogen

   c. Magnesium nitride with water

   d. Heating magnesium hydroxide

3B. Hydrates and Their Thermal Decompositions

Introduction

What is a *hydrate?* A hydrate is a salt that has crystallized from aqueous solution with weakly bound water molecules *contained* in the crystals (Ebbing/Gammon, Section 2.8).

Consider the crystallization of sodium acetate ($NaC_2H_3O_2$) as an example. This process is shown in Figure 3B.1. The crystals that you see in the photograph can be removed from the solution by filtration. If these crystals are then allowed to sit in the open air for a few hours, all of the residual moisture on their surfaces will evaporate. They will appear to be dry. Nevertheless, water will still be present, and it will be present in a definite stoichiometric amount. Chemical analysis would reveal that 3 water molecules accompany every formula unit of sodium acetate. This is shown by writing the formula as $NaC_2H_3O_2 \cdot 3H_2O$.

The definite stoichiometry occurs because the water in a hydrate occupies definite sites in the crystalline lattice, just as Na^+ and Cl^- ions occupy definite positions in the NaCl lattice. Because this water occupies definite sites, it must be present in definite stoichiometric amounts. The quantity of water will not change as long as the temperature (and pressure) is not altered significantly. A substantial increase in temperature, however, will cause the loss of hydrate water.

FIGURE 3B.1

The crystallization of the trihydrate of sodium acetate from an aqueous solution.

Purpose

You will use mass relationships to identify the product that you obtain from the thermal decomposition of barium chloride dihydrate ($BaCl_2 \cdot 2H_2O$). You will also have the opportunity to observe the thermal decomposition of several other hydrates.

Concept of the Experiment

In the first part of this experiment, you will examine the quantitative aspects of the thermal decomposition of $BaCl_2 \cdot 2H_2O$. When this substance is heated, there are at least three possible reactions:

$$BaCl_2 \cdot 2H_2O(s) \xrightarrow{\Delta} BaCl_2 \cdot H_2O(s) + H_2O(g)$$

$$BaCl_2 \cdot 2H_2O(s) \xrightarrow{\Delta} BaCl_2(s) + 2H_2O(g)$$

$$BaCl_2 \cdot 2H_2O(s) \xrightarrow{\Delta} BaO(s) + H_2O(g) + 2HCl(g)$$

The product of the first reaction is the monohydrate, $BaCl_2 \cdot H_2O$. Alternatively, all of the water may be lost so that anhydrous (water-free) $BaCl_2$ is obtained, as shown in the second equation. Finally, the decomposition may take the course shown in the third equation. A simple loss of water has not occurred. HCl is also liberated, so the oxide, BaO, remains when the reaction is complete.

How will you know which reaction has occurred? Because you will know the mass of the reactant, you will be able to calculate the mass of each possible product. You will then be able to identify the product by comparing the measured mass of the product with the masses that you have calculated. This reaction is particularly "well behaved," so identification of the product should be simpleif you have followed the directions carefully. Moreover, you will pool your results with those of your classmates to establish the precision of the experiment. You may do the calculations by hand, or, if your laboratory instructor wishes, you can use a computer and the Internet (Appendix A).

You will also have the opportunity to observe the colorful thermal decompositions of $CrCl_3 \cdot 6H_2O$, $CoCl_2 \cdot 6H_2O$, $NiSO_4 \cdot 7H_2O$, and $CuSO_4 \cdot 5H_2O$. Because these reactions are not as well behaved as the one involving $BaCl_2 \cdot 2H_2O$, you will study only their qualitative aspects. Remember that a reaction must accompany every color change. Simple loss of water may occur on gentle heating, but more extensive decomposition could occur with protracted heating at a high temperature (e.g. hottest part of the flame). For example, $CuSO_4 \cdot 5H_2O$ will lose water to give anhydrous $CuSO_4$, which, in turn, can lose gaseous SO_3 to give CuO. Moistened blue litmus paper would detect this reaction, however, because SO_3 reacts with water to give sulfuric acid (H_2SO_4), and acids cause blue litmus paper to turn pink.

Procedure

Getting Started

1. Obtain a crucible and lid. Wash, rinse, and dry them. Also obtain 4 small test tubes and 8 small strips of blue litmus paper.
2. Place the covered crucible in a clay triangle on an iron ring that is attached to a ring stand. Adjust the height of the ring so that the bottom of the crucible will be in the hottest part of the flame of your laboratory burner. Figure 3A.1 from Experiment 3A shows the correct arrangement.
3. Obtain directions for discarding the solid products that will be produced in this experiment from your laboratory instructor.

Doing the Quantitative Experiment

1. Heat the covered crucible in the hottest part of the flame for about 3 min. The bottom of the crucible should attain a red- hot glow during this time.
2. Move the burner and allow the crucible to cool (5–10 min). When the crucible is cool, you should feel no heat when you place one of your fingers about 1/2 inch from the bottom of the crucible.

CAUTION: Avoid burning your fingers. Do not touch the crucible or the iron ring at any time during this experiment.

3. When the covered crucible is cool, transfer it to the pan of your most precise balance. When moving the crucible to the balance, use crucible tongs and hold a wire gauze under the crucible. If you must wait to use the balance, do not place the crucible directly on the bench. Put it on the wire gauze or leave it in the clay triangle.
4. Obtain and record the mass of the covered crucible.
5. Repeat Steps 1 and 4 until two consecutive masses differ by not more than ±0.001 g (or any other range stipulated by your laboratory instructor). Use the last mass that you obtain in subsequent calculations.
6. Weigh a piece of waxed weighing paper and record its mass. Place the weighing paper onto the pan of the laboratory balance. Add about 1.4–1.5 g of $BaCl_2 \cdot 2H_2O$ to the waxed weighing paper.
7. Uncover the crucible using crucible tongs, add the sample of $BaCl_2 \cdot 2H_2O$ to the crucible, dispose of the weighing paper, and cover the crucible again.
8. Obtain and record the mass of the covered crucible and its contents. Calculate and record the mass of $BaCl_2 \cdot 2H_2O$ in the sample.
9. Transfer the crucible to the clay triangle, using tongs. Leave the lid slightly ajar.
10. Heat the crucible slowly by brushing it with the flame for about 2–3 min.
11. Heat the crucible in the hottest part of the flame for about 15 min. The bottom of the crucible should be red hot during this time.
12. Repeat Steps 2 through 5.
13. Calculate and record the mass of the product. Calculate the ratio

    Mass of product/mass of $BaCl_2 \cdot 2H_2O$

    Share your value with your classmates and obtain their values for this ratio.

Doing the Qualitative Experiment

1. Place a pea-sized portion of $CrCl_3 \cdot 6H_2O$ in one of the test tubes.
2. With a test tube clamp, hold the test tube at an angle of about 45°, and quickly move the test tube in and out of the hottest part of the flame from your burner.
3. Note and record all color changes. Immediately test with blue litmus paper the water that condenses on the upper surface of the test tube.
4. When no further color changes are apparent, hold the test tube directly in the hottest part of the flame for about 1 min. Repeat the test with moistened blue litmus paper. Record your observations.
5. Repeat Steps 1 through 4 three more times. First with pea-sized portions of $CoCl_2 \cdot 6H_2O$, then with $NiSO_4 \cdot 7H_2O$, and finally with $CuSO_4 \cdot 5H_2O$.

CAUTION: Before you leave the laboratory, make sure that your gas outlet and those of your neighbors are closed.

Date ____________ Student Name ____________
Course/Section ____________ Team Members ____________
Instructor ____________ ____________

Hydrates and Their Thermal Decompositions

Prelaboratory Assignment

1. Provide definitions for the following words:

   a. Hydrate - A salt that has crystallized from aqueous solution with weakly bound water molecules contained in crystals.

   b. Anhydrous - A substance containing no water.

2. What are the differences between water on the surface of the crystals of a hydrate and hydrate water itself?

   The water on the surface can be wiped off or will evaporate if the crystals are heated. In hydrate the water molecules are chemically bonded with the crystal molecules and cannot evaporate easily.

3. a. The thermal decomposition of 1.6608 g of $MgSO_4 \cdot H_2O$ produces 1.4446 g of a product in a well-behaved reaction. There are two possibilities:

$$MgSO_4 \cdot H_2O(s) \xrightarrow{\Delta} MgSO_4(s) + H_2O(g)$$

$$MgSO_4 \cdot H_2O(s) \xrightarrow{\Delta} MgO(s) + SO\ (g) + H\ O(g)$$

Complete the following table, and then identify the correct solid product by comparing the calculated masses of MgSO4 and MgO with the observed mass of the product.

Substance	Formula Weight	Moles	Mass (g)
$MgSO_4 \cdot H_2O$	138.38	0.012	1.6608
$MgSO_4$	120.37	0.012	0.3606
MgO	40.30	0.012	0.4836

Because the observed mass of the product is 1.4446 g, the product is $MgSO_4$.

b. What qualitative test could be used to substantiate this result?

$MgSO_4 \cdot H_2O(s) \rightarrow MgSO_4(s) + H_2O$

$Mg^{2+}(aq) + SO_4^{2-}(aq)$

4. What special safety precautions have been cited in this experiment?

Avoid burning finger. Do not touch the crucible or the iron ring at any time of the experiment. Make sure the gas outlet are closed.

Date ______________________ Student Name ______________________
Course/Section ______________________ Team Members ______________________
Instructor ______________________ ______________________

Hydrates and Their Thermal Decompositions

Results

1. *Thermal decomposition of $BaCl_2 \cdot 2H_2O$*

Mass of empty crucible and lid (g): 32.89 g ______________________

34.39 - 32.89 = 1.50 g
reactant

Mass of crucible, lid, and $BaCl_2 \cdot 2H_2O$ (g): 34.39

Mass of empty crucible and lid (g): 12.54 g

Mass of $BaCl_2 \cdot 2H_2O$ (g): 1.50 g

Mass of crucible, lid, and product (g): 34.14 ______________________

Mass of product (g): 1.25

Mass of product/mass of $BaCl_2 \cdot 2H_2O$: ______________________

Shared data (Include your own.)

2. *Observations on thermal decompositions of other hydrates*

$CrCl_3 \cdot 6H_2O$

$CrCl_3 + 6H_2O$

$CrCl_3 + 6H_2O + HCl$

$CrO + H_2O + HCl$

$CoCl_2 \cdot 6H_2O$ pink → $CoO + H_2O + HCl$

$NiSO_4 \cdot 7H_2O$ color blue

NO acid placed

$NiSO_4 + H_2O$

$CuSO_4 \cdot 5H_2O$

pink

$CoO + H_2SO_4$

Student name: ________________ Course/Section: __________________ Date: _______________

Questions

1. a. Compute the mean value of the ratio

$$\text{Mass of product/mass of } BaCl_2 \cdot 2H_2O$$

and its standard deviation using the shared data. Calculate the standard deviation using the technique described in Appendix A.

mass product

$n = (x_i - x)$

1.252

1.290

1.211

1.328

1.252

1.262

standarde deviation

0.045412767

b. Give a mean value for the ratio that is consistent with the precision of the data.

c. Use this value of the ratio and your own mass of the $BaCl_2 \cdot 2H_2O$ to calculate a mass of the product that is consistent with the precision of the experiment.

d. Complete the following table. Then identify the correct solid product by comparing the calculated masses of $BaCl_2 \cdot H_2O$, $BaCl_2$, and BaO with the observed mass of the product (the value calculated in Question 1c).

Substance	Formula Weight	Moles	Mass (g)
$BaCl_2 \cdot 2H_2O$	244.28 g/m	1.5 / 244.28 mol = 0.0060 mol	1.5 g
$BaCl_2 \cdot H_2O$	226.28 g mol	0.006 mole	0.006 x 226 g/mol = 1.36 g
$BaCl_2$	208.24 g moles	0.006	0.006 x 208.24 = 1.25 g
BaO	153.33 g	0.006 mole	0.006 mol x ... = .92 g

Because the observed mass of the product is 1.5 g, the product is $BaCl_2 \cdot 2H_2O$.

2. Write a balanced chemical equation that is consistent with each observation that you have made concerning the thermal decompositions of the other hydrates. More than one equation may be requir[ed] for each hydrate. Explain your reasoning for each equation.

4A. Conductivity of Aqueous Solutions

Introduction

Aqueous solutions of some substances will conduct electricity (Ebbing/Gammon, Section 4.1). These substances are called *electrolytes*. Substances whose solutions will not conduct an electric current are called *nonelectrolytes*.

When an electrolyte is dissolved in water, ions are produced. Ions are responsible for conductivity. We will distinguish between two types of substances that produce ions in solution. *Strong electrolytes* form aqueous solutions that conduct electricity very easily. The conductivity is high because strong electrolytes in solution exist almost entirely as ions. *Weak electrolytes*, on the other hand, form solutions that are poorer conductors. In solution, weak electrolytes exist primarily as molecular substances, with only a few ions.

Purpose

The conductivities of solutions (or the electric current that flows through these solutions) will allow you to distinguish among strong electrolytes, weak electrolytes, and nonelectrolytes. You will also be able to distinguish between strong acids and bases and weak acids and bases. Finally, you will have the opportunity to dilute an acid and a base to a desired concentration.

Concept of the Experiment

When a voltage (E) is applied across a conductor, the current (i) is controlled by the resistance (R) of the conductor. The exact relationship is given by Ohm's law, $E = iR$. This equation can be rearranged to give an equivalent relationship, $1/R = i/E$. The term $1/R$ is called the *conductance*; it has units of ohm^{-1}. This equation also shows that the conductance is directly proportional to the current at a constant voltage. Therefore, a measurement of the current is an indirect measurement of the conductance. Some laboratories may not have a conductivity apparatus. Therefore, do not be surprised if you are asked to measure current rather than conductance in this experiment.

The magnitude of each conductance (or current) that you measure will depend on the concentration of the ions in the solution and on their type. These two factors will dictate a pattern of conductances that will enable you to pick out the strong electrolytes, the weak electrolytes, and the nonelectrolytes.

Do not expect, however, that *isomolar* solutions (solutions that have the same molarity) of different strong electrolytes will have the same conductance. Although each solution will have a large conductance, each conductance will differ from the others. The reason lies in the differing mobilities of different ions. The abilities of different ions to move through the solution will not be identical. As a result, their conductances will differ even though their concentrations are the same.

Finally, we will note the existence of an *approximate* additivity rule. As an example, consider a dilute solution containing two strong electrolytes. We will stipulate that these substances do not react with each other. Consider, also, separate solutions of these electrolytes. The concentration of each electrolyte in its separate solution must be identical to its concentration in the mixture. Under these conditions, the conductance of the mixture will be approximately equal to the sum of the conductances of the separate solutions.

Additivity will not occur, however, when the two electrolytes react with each other. The reason is easy to understand. A chemical reaction will produce new substances with different characteristic conductances. In this experiment, a noticably large deviation from additivity will allow you to pinpoint a case in which a chemical reaction has occurred.

A Special Note About Molarity and Dilution

In this experiment, you will encounter the symbol *M*. It stands for molarity and has units of mol/L (Ebbing/Gammon, Section 4.7). Molarity is a measure of the concentration of a solution.

You will also encounter dilution—perhaps for the first time. Suppose you dilute (by adding water) a solution of known initial molarity (M_i) and known initial volume (V_i) to a solution with a final molarity (M_f) and final volume (V_f). The following equation (Ebbing/Gammon, Section 4.8) is applicable:

$$M_i \times V_i = M_f \times V_f$$

In this experiment, you will know M_i, M_f, and V_f. You will be able to calculate V_i from this equation. The difference, $V_f - V_i$, will tell you how much water to add to achieve the desired dilution (if it is assumed that the volume of the original solution and the volume of the water added are perfectly additive).

Procedure

Getting Started

1. Your laboratory instructor may ask you to work with a partner.
2. Obtain an apparatus to measure either conductance or current.

   CAUTION: Handle the apparatus with care to avoid electric shocks.

3. Prepare 0.10 *M* solutions of acetic acid ($HC_2H_3O_2$) and ammonia (NH_3) according to the directions you devised in the Prelaboratory Assignment.

   CAUTION: Handle the 6 *M* solutions of acetic acid and ammonia with caution. Wear approved chemical splash goggles. Work in an area with very good ventilation. A hood is preferred. Do not mix these substances with each other. If you spill one of these substances on you, wash the contaminated area thoroughly and report the incident to your laboratory instructor. You may need further treatment. Spills elsewhere (floor, bench-top) should be reported as well.

Making the Measurements

1. Use a 50-mL beaker for each of the following measurements. Rinse it and dry it after each measurement.
2. Compare the conductances of 20 mL of distilled water, 20 mL of tap water, and 20 mL of ethyl alcohol (C_2H_5OH). Measure the volumes of each of these liquids and then transfer the liquids to the 50 mL beaker.. Record the results from each conductance measurement.
3. Measure the conductances of 0.10 *M*, 0.050 *M*, and 0.020 *M* solutions of HCl. These solutions should be prepared according to the directions you outlined in the Prelaboratory Assignment. Make sure that each solution is mixed thoroughly before you measure the conductance.

4. Add 10 mL of 0.10 M HNO_3 to the beaker from the clean, dry graduated cylinder. Add 10 mL of distilled water. Swirl gently to mix. Calculate and record the new concentration. Measure and record the conductance.

5. Repeat Step 4, substituting, in turn (i.e. one solution at a time), 0.10 M KOH, 0.10 M KCl, 0.10 M KNO_3, and 0.10 M $Ca(NO_3)_2$ for 0.10 M HNO_3.

6. Repeat Step 4, substituting, in turn, your 0.10 M NH_3 and your 0.10 M $HC_2H_3O_2$ for 0.10 M HNO_3.

7. Measure and record the conductance of a solution containing 10 mL of 0.10 M HCl and 10 mL of 0.10 M KNO_3.

8. Measure and record the conductance of a solution containing 10 mL of 0.10 M HNO_3 and 10 mL of 0.10 M KCl.

9. Measure and record the conductance of a solution containing 10 mL of 0.10 M NH_3 and 10 mL of 0.10 M $HC_2H_3O_2$.

10. Calculate and record the new concentration of each substance in Steps 7, 8, and 9.

Date ____________ Student Name ____________
Course/Section ____________ Team Members ____________
Instructor ____________ ____________

Conductivity of Aqueous Solutions

Prelaboratory Assignment

1. Define the following terms:

   a. Electrolyte

   b. Strong electrolyte

   c. Weak electrolyte

   d. Nonelectrolyte

2. What is the relationship, if any, between the following pairs of substances (see Ebbing/Gammon, Section 4.4)?

   a. Strong electrolyte and strong acid or base

b. Weak electrolyte and weak acid or base

3. Identify the acids and the bases used in this experiment (see Ebbing/Gammon, Section 4.4).

4. a. Describe, step-by-step, how will you prepare 10 mL of 0.050 *M* HCl from a 0.10 *M* solution of HCl? Assume that you use a 10-mL graduated cylinder (in which volumes can be read to the nearest 0.1 mL).

b. Describe, step-by-step, how will you prepare 10 mL of 0.020 *M* HCl from the 0.050 *M* solution?

Student name: ______________ Course/Section: ______________ Date: ______________

5. How will you prepare the following solutions from 6.0 *M* solutions of acetic acid and ammonia? Give exact details. Assume that a 10-mL graduated cylinder (in which volumes can be read to the nearest 0.1 mL) and a 100-mL graduated cylinder are available.

   a. 80 mL of 0.10 *M* $HC_2H_3O_2$

   b. 80 mL of 0.10 *M* NH_3

6. What special safety precautions must be observed during this experiment?

Date ____________________ Student Name ____________________
Course/Section ____________________ Team Members ____________________
Instructor ____________________ ____________________

Conductivity of Aqueous Solutions

Results

Substance or Solution	Concentration (*M*)	Conductance or Current
Distilled water	X	
Tap water	X	
C_2H_5OH	X	
HCl	0.10	
HCl	0.050	
HCl	0.020	
HNO_3		
KOH		
KCl		
KNO_3		
$Ca(NO_3)_2$		
NH_3		
$HC_2H_3O_2$		

Mixture		Concentration (M)	Conductance or Current
$HCl + KNO_3$	HCl	________	________
	KNO_3	________	
$HNO_3 + KCl$	HNO_3	________	________
	KCl	________	
$HC_2H_3O_2 + NH_3$	$HC_2H_3O_2$	________	________
	NH_3	________	

Student name: ______________ Course/Section: ______________ Date: ______________

Questions

1. a. Use your measured conductances (currents) for isomolar solutions of the substances you tested and the conductances (currents) for pure water and ethyl alcohol to distinguish among strong electrolytes (SE), weak electrolytes (WE), and nonelectrolytes (NE).

H_2O	______________	KNO_3	______________
C_2H_5OH	______________	KOH	______________
HCl	______________	$Ca(NO_3)_2$	______________
HNO_3	______________	$HC_2H_3O_2$	______________
KCl	______________	NH_3	______________

b. Use these results to classify the acids and bases as strong acids (SA), weak acids (WA), strong bases (SB), or weak bases (WB).

HCl	______________	KOH	______________
HNO_3	______________	NH_3	______________
$HC_2H_3O_2$	______________		

2. Use your experimental results to offer explanations, where possible, for the following observations.

a. The effect of the concentration of an electrolyte on the conductance (current)

b. The cause of the differences in the conductances (currents) of isomolar solutions of KNO_3 and $Ca(NO_3)_2$

c. The presence of ionic impurities in ordinary tap water

d. The presence of molecular impurities in ordinary tap water

3. For each of the mixtures of electrolytes, compare the conductance (current) of the mixture with the sum of the conductances (currents) of the separate electrolytes. If the additivity rule does not appear to be true, offer an explanation for the increased or decreased conductance (current) and give an equation for the reaction. It may be helpful for you to read the section on neutralization in your textbook (Ebbing/Gammon, Section 4.4) before you formulate your answer.

4B. Ionic Reactions in Aqueous Solutions

Introduction

Reactions in aqueous solutions (Ebbing/Gammon, Chapter 4) have far-reaching importance. These reactions occur in our homes as well as in rivers, lakes, and oceans; in biological systems such as our bodies; and in many industrial applications. Most of these reactions involve ions.

Purpose

You will examine precipitation reactions and reactions of acids and bases.

Concept of the Experiment

You will have an opportunity to examine certain precipitation reactions and test the solubility rules (Ebbing/Gammon, Section 4.1) shown in Table 4B.1. If you did Experiment 1A, "Identification of an Unknown Compound," you will already have some familiarity with the formation of precipitates.

Table 4B.1

Empirical Rules for the Solubilities of Common Ionic Compounds

Soluble Compounds	Exceptions
Sodium, potassium, and ammonium compounds	None
Acetates and nitrates	None
Chlorides	Lead and silver chlorides are insoluble.
Sulfates	Calcium, barium, and lead(II) sulfates are insoluble.
Insoluble Compounds	**Exceptions**
Carbonates and phosphates	Sodium, potassium, and ammonium compounds are soluble.
Hydroxides	Sodium, potassium, and calcium compounds are soluble.
Sulfides	Sodium, potassium, calcium, and ammonium compounds are soluble.

You will also examine a reaction in which a gas is formed and some reactions of acids and bases (Ebbing/Gammon, Section 4.4). You will find that H^+ ions (protons) from acids cause blue litmus paper to turn red or pink. Similarly, you will see that OH^- (hydroxide) ions from bases cause pink litmus paper to turn blue. Moreover, you will find that the release of heat (a signal of a chemical reaction) accompanies the reaction of an acid with a base.

A Word About Molarity

In this experiment, you will encounter the symbol *M*. It stands for *molarity* and has units of moles per liter (mol/L) (Ebbing/Gammon, Section 4.7). It also appeared in Experiment 4A. Molarity is a measure of the concentration of a solution.

Procedure

Getting Started

1. Obtain several pieces of red and blue litmus paper and 4 small test tubes.
2. Use a 5-mL or a 10-mL graduated cylinder to place 1 mL of distilled water in each of these test tubes. Mark the height of the water with a marking pencil or a small piece of tape. Add an additional 1 mL to each of these test tubes, and mark the new height of the water. Pour the water out of each test tube.
3. Obtain instructions for using the centrifuges from your laboratory instructor.

   CAUTION: When you use a centrifuge, do not attempt to stop the centrifuge rotor with your finger or anything else.

4. Obtain directions from your laboratory instructor for discarding the solutions that you use during this experiment.
5. Remember to be careful in your handling of the solutions in this experiment.

   CAUTION: Sodium hydroxide, hydrochloric acid, acetic acid, and ammonia can cause chemical burns, in addition to ruining your clothes. If you spill any of these solutions on you, wash the contaminated area thoroughly with water, and immediately report the incident to your instructor. You may require further treatment.

Testing the Solubility Rules

1. Using the lower marks on the 4 test tubes as guides, add 1 mL of 0.1 M NH_4NO_3 to each test tube.
2. Using the upper marks as guides, add 1 mL of 0.1 M NaBr to the first tube, 1 mL of 0.1 M Na_2SO_4 to the second, 1 mL of 2 M NaOH to the third, and 1 mL of 0.1 M Na_2CO_3 to the fourth. Shake each test tube gently. Record your observations, noting the colors of all precipitates.
3. Discard the contents of each test tube as directed by your laboratory instructor. Wash the test tubes carefully, and rinse them with distilled water.
4. Repeat Steps 1 through 3 with, in turn, 0.1 M $Ba(NO_3)_2$, 0.1 M $AgNO_3$, 0.1 M $Pb(NO_3)_2$, and 0.1 M $Ni(NO_3)_2$ instead of NH_4NO_3 (see Step 5).

   CAUTION: Wash your hands thoroughly after using the solution containing barium, because it is poisonous.

5. Centrifuge the contents of the test tube that originally contained 0.1 M $Ni(NO_3)_2$ and 2 M NaOH. About 1 min will be required. Decant (pour off) and discard the solution. Save the precipitate for use in step 3 of the next part of this experiment: "Looking at Acids and Bases".

Looking at Acids and Bases

1. Wash the three remaining test tubes, and rinse them with distilled water.
2. Obtain a pea-sized portion of $CaCO_3$ in one of the test tubes. Add 20 drops of 2 M HCl. Record your observations.
3. Add 8 drops of 2 M HCl to the precipitate that you saved from the reaction between $Ni(NO_3)_2$ and NaOH. Record your observations.

4. Discard the contents of the test tubes, wash each test tube carefullym, and rinse them with distilled water.
5. Using the lower marks as guides, add 1 mL of 2 *M* HCl to one test tube, 1 mL of 2 *M* $HC_2H_3O_2$ (acetic acid) to the second, 1 mL of 2 *M* NH_3 to the third, and 1 mL of 2 *M* NaOH to the fourth.
6. Take a drop of each solution using a clean stirring rod, and touch it to a piece of red litmus paper. Record your observations.
7. Repeat Step 6 using blue litmus paper.
8. Add the contents of the test tube containing HCl to the test tube containing NH_3. Does the test tube feel warmer after the HCl is added? If the test tube feels warmer, then heat is *evolved* or given off when the chemical reaction takes place.Add the contents of the test tube containing $HC_2H_3O_2$ to the test tube containing NaOH. Is heat evolved? Record the results.

Date ____________________ Student Name ____________________
Course/Section ____________________ Team Members ____________________
Instructor ____________________ ____________________

Ionic Reactions in Aqueous Solutions

Prelaboratory Assignment

1. Provide definitions for the following terms:

   a. Exchange (metathesis) reaction

   b. Acid

   c. Base

   d. Neutralization

2. Predict the solubilities of the following substances in water, using Table 4B.1. These substances are relevant to this experiment.

a. $NaNO_3$

b. NH_4Br

c. $(NH_4)_2SO_4$

d. $(NH_4)_2CO_3$

e. $BaBr_2$

f. $BaSO_4$

g. $Ba(OH)_2$

h. $BaCO_3$

3. Give names and formulas for the acids and bases that you will encounter in this experiment.

4. What safety precautions must be observed during this experiment?

Date		Student Name	
Course/Section		Team Members	
Instructor			

Ionic Reactions in Aqueous Solutions

Results

1. *Testing the solubility rules*

	$NaBr$	Na_2SO_4	$NaOH$	Na_2CO_3
NH_4NO_3				
$Ba(NO_3)_2$				
$AgNO_3$				
$Pb(NO_3)_2$				
$Ni(NO_3)_2$				

2. *Looking at acids and base*

$HCl + CaCO_3$		**HCl + precipitate from $Ni(NO_3)_2$ + NaOH**	

	Red Litmus Paper	**Blue Litmus Paper**
HCl		
$HC_2H_3O_2$		
NH_3		
NaOH		

	Heat Evolved?
$HCl + NH_3$	
$HC_2H_3O_2 + NaOH$	

Student name: ______________ Course/Section: ________________ Date: ______________

Questions

1. a. Write balanced molecular and net ionic equations for each successful precipitation reaction that you observed.

b. Were the solubility rules completely adequate? Explain.

2. a. Describe how red litmus paper is affected by acids and bases.

b. Describe how blue litmus paper is affected by acids and bases.

c. Which litmus paper would you use to test for an acid? Why?

d. Which litmus paper would you use to test for a base? Why?

3. a. Write balanced molecular and net ionic equations for each reaction of a solid with HCl.

b. Write balanced molecular and net ionic equations for each neutralization reaction in which heat was evolved.

4C. How Much Acetic Acid Is in Vinegar?

Introduction

Quantitative analysis determines the amount of a particular substance in a sample. Often, this determination is accomplished through a *titration* (Ebbing/Gammon, Section 4.10) of the sample with a solution of another substance whose concentration is known. Of course, these substances must react with each other.

Some physical or chemical response must be coupled with the titration to signal the exact completion of the reaction. An *indicator* can sometimes be used. The indicator chosen will have one color before the reaction is complete and another color when completion occurs. After the indicator is added to the sample solution, the solution of known concentration is delivered carefully from a buret until the indicator changes color. Before performing this experiment, you should review the proper use of a buret which is described in the Introduction to this manual.

Purpose

You will determine the amount of acetic acid in white vinegar by titration with a solution of sodium hydroxide whose concentration is known. The indicator will be phenolphthalein.

Concept of the Experiment

The object of this experiment is to determine the molar concentration (Ebbing/Gammon, Section 4.7) of acetic acid ($HC_2H_3O_2$) in vinegar. You will accomplish this through the titration of a sample of vinegar with a solution of the base sodium hydroxide (NaOH). These substances react quickly and completely.

$$HC_2H_3O_2 + NaOH \rightarrow NaC_2H_3O_2 + H_2O$$

Phenolphthalein will be used as the indicator. It will be colorless before the completion of this reaction but pink after completion. You must be prepared to search carefully for a point in the titration at which 1 drop of the NaOH solution causes the solution being titrated to turn from colorless to a barely discernible pink color. This point is called the *endpoint.* You will do a trial titration to find the approximate endpoint before you do a pair of exact titrations. You will be able to determine the molarity of the acetic acid in the vinegar using the molarity of the solution of sodium hydroxide, the volume of sodium hydroxide needed to reach the endpoint of the titration and the balanced equation shown above.

Procedure

Getting Started

1. Obtain a 10-mL transfer pipet and a 50-mL buret.
2. Next obtain about 30 mL of white 5% vinegar and about 85 mL of the solution of NaOH. The vinegar may be kept in a clean, dry beaker. However, the NaOH solution must be kept in a clean, dry Erlenmeyer flask that is closed with a rubber stopper. This solution must be protected when it is not in use because NaOH will react with carbon dioxide (CO_2) in the air. The molarity of this solution will lie between 0.2 *M* and 0.3 *M*; the exact concentration will be given on the label of the bottle.

Cleaning and Filling Your Buret

1. The Introduction to this manual gives instructions for using a buret. Clean your buret and fill it with the NaOH solution after you have read those procedures carefully.

Doing the Trial Titration

1. Pipet 10.0 mL of vinegar into a clean 125-mL Erlenmeyer flask. Add about 20 mL of distilled water from a clean graduated cylinder. Add 2 drops of phenolphthalein solution.

   CAUTION: Never use your mouth to draw liquid into the pipet. Use a rubber suction bulb or some other suction device.

2. Record the molarity of the NaOH solution and the initial buret volume to two decimal places
3. Place the flask under the buret with the capillary tip inside the mouth of the flask. Place a piece of white paper under the flask.
4. Add the NaOH solution to the flask in increments of about 1 mL while swirling solution in the flask. Note the color of the solution after each addition.
5. This trial titration is complete when an addition of about 1 mL causes the color to change from colorless to any shade of pink.
6. Record the final buret volume to two decimal places. Subtract the initial volume from the final volume to obtain the volume of sodium hydroxide required to reach the approximate endpoint.

Doing the Exact Titrations

1. Repeat Steps 1, 2, and 3 of the procedure used for the trial titration.
2. Subtract 1 mL from the volume found in the trial titration. *Rapidly* add the resulting volume to the flask from the buret.
3. Rinse the walls of the flask with distilled water from a plastic wash bottle.
4. Continue the titration on a *drop-by-drop* basis. Swirl the flask rapidly after each drop. The endpoint is the first permanent, *barely visible* pink color. Finding the true endpoint requires patience and skill. Absolutely no skill is required to miss the endpoint and achieve a very deep pink color. If you are unsure about the endpoint, record the buret volume before you add the next drop.
5. Repeat the procedure with a second sample of vinegar.
6. If the volumes at the endpoints of these two exact titrations differ by more than 0.15 mL (about 3 drops), repeat the titrations with additional samples of vinegar until two consecutive results have this precision.
7. Calculate and record the molarity of the vinegar from each of the two titrations. Does each molarity have the correct number of significant figures? Obtain the mean molarity.

Date ______________________ Student Name ______________________
Course/Section ______________________ Team Members ______________________
Instructor ______________________ ______________________

How Much Acetic Acid Is in Vinegar?

Prelaboratory Assignment

1. a. What is the purpose of the trial titration?

   b. Describe the *endpoint* of a titration. What is it? How will you know when you have reached it?.

   c. Explain why the walls of the flask are washed with distilled water during the titration. How will this affect the outcome of the titration?

2. A 10.0-mL sample of aqueous HCl requires 32.34 mL of 0.108 *M* NaOH to reach the endpoint. What is the molar concentration of HCl? The equation for the reaction is

$$NaOH + HCl \rightarrow NaCl + H_2O$$

3. What safety rule must be observed during this experiment?

Date	____________	Student Name	____________
Course/Section	____________	Team Members	____________
Instructor	____________		____________

How Much Acetic Acid Is in Vinegar?

Results

Molarity of the NaOH solution: ____________

1. *Trial titration*
   Final buret volume (mL): ____________

   Initial buret volume (mL): ____________

   Volume of NaOH solution (mL): ____________

2. *Exact titrations*

Sample No.	**1**	**2**	**3**	**4**
Final buret volume (mL)	______	______	______	______
Initial buret volume (mL)	______	______	______	______
Volume of NaOH solution (mL)	______	______	______	______
Concentration of $HC_2H_3O_2$ (*M*)	______	______	______	______
Mean Concentration (*M*)	______	______	______	______

Calculations:

Questions

1. The manufacturer of the vinegar used in this experiment claims that the vinegar contains 5% acetic acid by weight. Use your results and a density of 1.0 g/mL to determine if this claim is true or false.

2. Repeat the calculation for problem 2 in the Prelaboratory Assignment, substituting H_2SO_4 for HCl.

5. The Decomposition of Potassium Chlorate

Introduction

Small quantities of molecular oxygen (O_2) can be obtained from the thermal decomposition of various oxygen-containing compounds. Some examples of these reactions are

$$2Ag_2O(s) \xrightarrow{\Delta} 4Ag(s) + O_2(g)$$

$$2BaO_2(s) \xrightarrow{\Delta} 2BaO(s) + O_2(g)$$

$$2KClO_3(s) \xrightarrow{\Delta} 2KCl(s) + 3\ O_2(g)$$

The last reaction includes manganese dioxide (MnO_2) as a *catalyst*. This reaction is particularly important in this experiment.

Purpose

You will study the thermal decomposition of $KClO_3$ (potassium chlorate) in the absence of a catalyst. You will be able to identify the solid that remains after the decomposition from the quantity of oxygen that is produced. You will verify this identification by comparing the measured mass of the solid product with a calculated value. You will determine if the solid is KCl, or if the absence of a catalyst affected the identity of the products of the reaction.

A Word About Catalysts

A catalyst is a substance that causes an increase in the rate of a chemical reaction without being used up in the reaction. Moreover, a catalyst does not *usually* alter the end result of a reaction. The same products are usually obtained in the presence of a catalyst and in its absence.

Concept of the Experiment

The solid product that results from the thermal decomposition of $KClO_3$ is KCl when the catalyst MnO_2 is present. What do you think will happen if the catalyst is not added? Will the loss of oxygen be less extensive? If so, the solid product will be either $KClO_2$ (potassium chlorite) or KClO (potassium hypochlorite). How do you think this experiment will answer these questions?

You will heat a sample of $KClO_3$ of known mass in the absence of a catalyst until the evolution of oxygen is complete. Oxygen will be collected in a flask by the displacement of water (Ebbing/Gammon, Section 5.5). In order to determine the correct stoichiometry of this reaction, you will need to obtain the number of moles of O_2 that have been evolved. You can calculate this quantity from the rearranged form of the ideal gas law, $n = PV/RT$. Clearly, you will need to measure the pressure, volume, and temperature of the oxygen.

Because the oxygen is collected by displacing water, water vapor will also be present in the gas. The experiment is designed so that the total pressure of the oxygen and water vapor will be equal to the atmospheric pressure; you can easily measure this quantity with a barometer. The partial pressure of oxygen in the flask can be calculated from the total pressure and the vapor pressure of water. Table 5B.1 gives the vapor pressure of water at various temperatures.

Figure 5B.1 shows a suitable apparatus for this experiment. The sample of $KClO_3$ is placed in the test tube, and the flask is filled with water. Some of the water is displaced by oxygen and is pushed into the beaker. The volume of water in the beaker will be identical to the volume of oxygen in the flask.

Procedure

Getting Started

1. Your laboratory instructor may ask you to work with a partner.
2. Obtain and clean the glassware for the apparatus in this experiment.
3. Obtain a wooden splint (a small piece of wood).
4. Obtain and record the atmospheric pressure in the laboratory.
5. Obtain directions from your laboratory instructor for discarding the solid product that will be produced in this experiment.
6. *Without fail*, observe the following safety precaution throughout this experiment:

   CAUTION: $KClO_3$ is a very strong oxidizing agent. Do not under any circumstances let this substance contact paper or the rubber stopper in the test tube belonging to the apparatus. A fire or explosion can occur.

Setting Up the Apparatus

1. Assemble the apparatus as shown in Figure 5B.1.
2. Test the apparatus for leaks, using the method in Steps 3 through 6. Remember that a leak will ruin your results.
3. Fill the flask almost completely with water, and open the pinch clamp.
4. Remove the stopper from the test tube. Using a large pipet bulb, push air through the glass tube until the rubber tube is filled with water. Water will begin to flow from the flask to the beaker.
5. Replace the test tube while the water is flowing.
6. If no leak is present, no more water will flow and the water level in the flask will no longer change. If a leak is present, water will continue to flow.
7. If a leak is indicated by this test, it is almost certainly located around one of the rubber stoppers.

   Press each one firmly into place. If water continues to flow, ask your laboratory instructor for assistance.
8. Close the pinch clamp. Water should stay in the rubber tube. Do not drain it or try to shake it out.

FIGURE 5B.1

A suitable apparatus for the thermal decomposition of $KClO_3$.

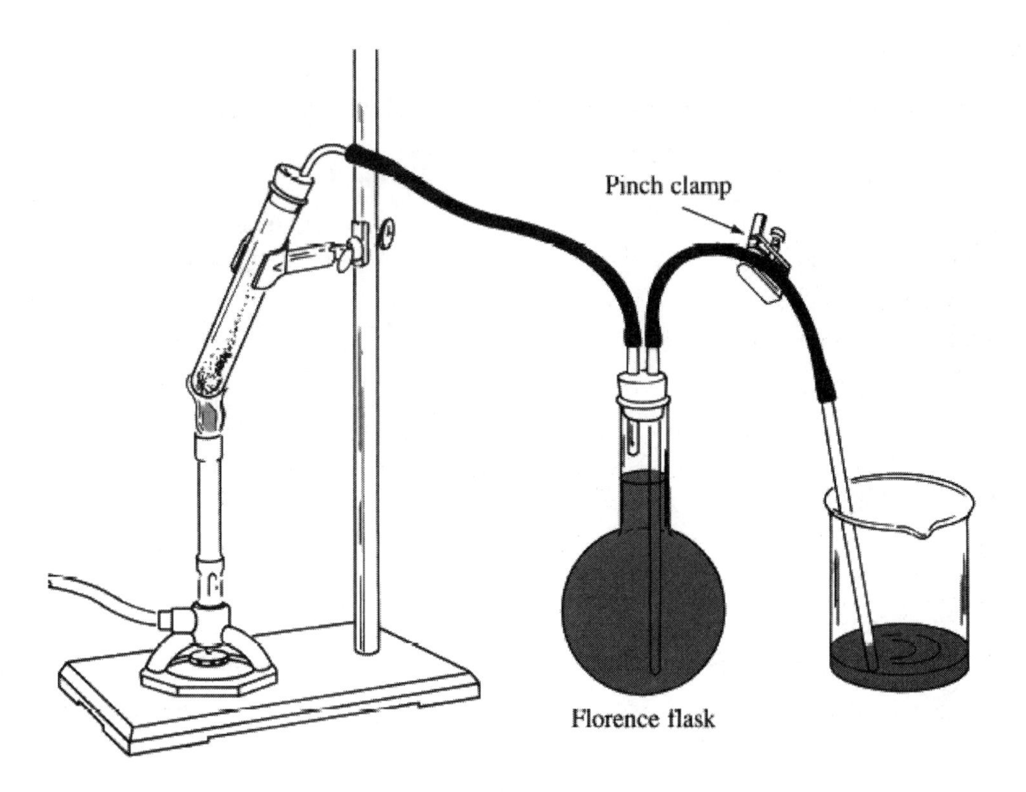

Doing the Experiment

1. Make sure that the test tube is clean and dry. Measure and record its mass to the nearest 0.001 g (or any other precision stipulated by your laboratory instructor).
2. Your laboratory instructor will put about 1.0 g of $KClO_3$ into the test tube.

   CAUTION: Care should be taken when adding potassium chlorate to the test tube to keep the amount at 1 gram or less. Adding too much material to the test tube will result in a buildup of gas pressure that could result in the shattering of the test tube.

3. Measure and record the mass of the test tube and its contents with the same precision as in Step 1.

   Calculate and record the mass of $KClO_3$.

4. Do not connect the test tube to the apparatus yet.
5. Fill the beaker with water. Open the clamp and raise the beaker above the flask. Water will siphon back into the flask. Allow the water in the flask to rise until it is about 1 inch below the glass tube leading to the test tube. Close the pinch clamp.
6. Connect the test tube containing $KClO_3$ to its rubber stopper. None of the solid material should touch the rubber stopper or even be near it.
7. Equalize the pressure in the apparatus and atmospheric pressure. To do this, you must open the pinch clamp and adjust the height of the beaker until the water levels in the flask and in the beaker are identical. Close the clamp.

8. Empty the beaker, but do not dry it. The volume of the water drops that remain in the beaker will be roughly equal to the volume that will remain after the displaced water is poured into a graduated cylinder for measurement. Do not empty or shake out the rubber tube when you empty the beaker.

9. Place the rubber tube back in the beaker and *open the clamp*.

CAUTION: If the clamp is not opened at this point, the buildup of gas could cause an explosion, although it is more likely that a stopper would be forced to loosen.

10. Use your laboratory burner to heat the test tube. Be cautious at first and brush the flame over the test tube. After a few minutes, you can heat the test tube more strongly by having the flame under the test tube for a longer period of time. The solid will melt, oxygen will be evolved, and water from the flask will be displaced into the beaker.

11. Heat until no more gas is produced and water from the flask no longer is pushed into the beaker. The contents of the test tube may solidify, because the melting point of the product is greater than that of $KClO_3$.

12. Remove the flame, and allow the system to come back to room temperature. About 15–20 min will be required.

13. Equalize the pressure, using the same procedure as in Step 7. The total pressure of the gas (oxygen plus water vapor) will then be equal to the atmospheric pressure.

14. With the clamp closed, remove the tube from the beaker. Measure the volume of water in the beaker by pouring it into a 1-L graduated cylinder. Record the volume.

15. Measure the temperature of the water to the nearest degree. This temperature will be taken as the temperature of the gas. You should also use it to determine the appropriate vapor pressure of water from Table 5B.1.

Table 5B.1

Vapor Pressure of Water

Temperature (°C)	Vapor Pressure (mmHg)	Temperature (°C)	Vapor Pressure (mmHg)
20	17.5	25	23.7
21	18.6	26	25.2
22	19.8	27	26.7
23	21.1	28	28.3
24	22.4	29	30.0

16. Obtain the mass of the test tube and its contents. Calculate and record the mass of the product.

17. Ignite the wooden splint, let it burn for a short time, and then blow out the flame. A glowing ember should remain. Thrust the glowing end of the splint into the flask, and observe the results. Record your observation.

18. If time permits, repeat Steps 1 through 16 with a clean test tube and a fresh sample of $KClO_3$.

CAUTION: Before you leave the laboratory, make sure that your gas outlet and those of your neighbors are closed.

Date	______________________	Student Name	______________________
Course/Section	______________________	Team Members	______________________
Instructor	______________________		______________________

The Decomposition of Potassium Chlorate

Prelaboratory Assignment

1. How will you measure the volume of the oxygen in this experiment?

2. A sample of helium is collected by displacing 182 mL of water. The pressures of the gas and the atmosphere are equalized, using the method described in this experiment. The atmospheric pressure is 738 mmHg. The temperature of the water and that of the gas are identical at 26°C. How many moles of helium have been evolved?

3. What safety precautions are cited in this experiment?

Date		Student Name	
Course/Section		Team Members	
Instructor			

The Decomposition of Potassium Chlorate

Results

Atmospheric pressure: ____________ mmHg

Trial	1	2
Mass of test tube + $KClO_3$ (g)		
Mass of empty test tube (g)		
Mass of $KClO_3$ (g)		
Volume of water displaced (mL)		
Temperature of water (°C)		
Mass of test tube + product (g)		
Mass of empty test tube (g)		
Mass of product (g)		

Observation with the glowing splint:

Questions

1. Using the data that you have accumulated, complete the following table. Show your calculations.

Trial	1	2
Pressure of gas (mmHg)	____________	____________
Vapor pressure of water (mmHg)	____________	____________
Pressure of O_2 (mmHg)	____________	____________
Volume of O_2 (mL)	____________	____________
Temperature of O_2 (°C)	____________	____________
Moles of O_2	____________	____________
Moles of $KClO_3$ in test tube	____________	____________
Moles of O_2/mol $KClO_3$	____________	____________
Mean moles of O_2/mol $KClO_3$	____________	____________

Calculations:

Student name: ________________ Course/Section: __________________ Date: ________________

2. a. Write a balanced chemical equation for each possible product in this experiment.

b. Using the masses of $KClO_3$ that you placed into the test tube for this experiment, calculate the expected masses of each solid product from these equations.

3. a. Based on the ratio of mean moles of O_2/mol $KClO_3$, what is the chemical equation for the reaction that occurred,? Why? Is this conclusion in accord with the mass of the solid product that you obtained? Explain.

b. Did the absence of the catalyst have and effect on the outcome of the thermal decomposition of $KClO_3$? What data supports this conclusion?

4. What does your observation with the glowing splint indicate about the nature and reactivity of oxygen?

6. Thermochemistry and Hess's Law

Introduction

The energy changes that accompany chemical reactions are nearly always reflected by the release or absorption of heat (q). There are many practical and theoretical reasons for studying the quantitative aspects of this phenomenon. Each of these studies is an application of *thermochemistry* (Ebbing/Gammon, Chapter 6).

Purpose

In this experiment you and your classmates will measure the enthalpy change, ΔH, that occurs during a chemical reaction. The enthalpy change is the amount of heat absorbed or produced during a chemical reaction at a constant pressure (e.g. atmospheric pressure). One-half of your class will measure the enthalpy change that occurs when solutions of sodium hydroxide (NaOH) and hydrochloric acid (HCl) are mixed. The other half will measure the enthalpy change that occurs when solutions of ammonia (NH_3) and hydrochloric acid are mixed. Both groups will measure the enthalpy change that occurs when solutions of sodium hydroxide and ammonium chloride (NH_4Cl) are mixed. All of the data will be pooled. The mean enthalpy changes from the first two measurements will enable you to determine the agreement between the mean enthalpy change from the third measurement and that predicted by Hess's law.

Concept of the Experiment

The chemical reactions that will occur during this experiment are given in the following equations:

$$NaOH + HCl \rightarrow NaCl + H_2O$$

$$NH_3 + HCl \rightarrow NH_4Cl$$

$$NaOH + NH_4Cl \rightarrow NaCl + NH_3 + H_2O$$

You may have noted that there is a relationship among these equations. If you reverse the second equation and add it to the first equation, the third equation is generated. This relationship provides the basis for using Hess's law of heat summation (Ebbing/Gammon, Section 6.7). You will find that you can predict the enthalpy change for the third reaction by combining the enthalpy changes for the first and second reactions.

The heat evolved or absorbed during these reactions will be measured with a coffee-cup calorimeter, whose use is described in Appendix B. This appendix also includes pertinent definitions of the system and its surroundings (Ebbing/Gammon, Section 6.2) in terms of this calorimeter. The three chemical reactions that will occur in this experiment are examples of bases reacting with acids. Although you will learn much more about these terms during this course, it will be useful for us at least to identify the bases and the acids in this experiment.

Bases: NaOH and NH_3

Acids: HCl and NH_4Cl

A Special Note

By sharing your data with your classmates, you eliminate the need to observe all of the reactions yourself. Moreover, the accuracy of the measured enthalpy changes will be increased.

Procedure

Getting Started

1. Work with a partner.
2. Obtain a coffee-cup calorimeter.
3. Ask your laboratory instructor which enthalpy change (NaOH–HCl or NH_3–HCl) you should measure.

Measuring the Evolution or Absorption of Heat

1. Place exactly 50 mL of the 2.0 *M* solution of HCl in a clean, dry graduated cylinder. Place exactly 50 mL of the 2.0 *M* solution of either NaOH or NH_3 in another clean, dry graduated cylinder.

   CAUTION: Hydrochloric acid, sodium hydroxide, and ammonia can cause chemical burns, in addition to ruining your clothing. If you spill one of these solutions on you, wash the contaminated area thoroughly and report the incident to your laboratory instructor. You may require further treatment. Always wear approved chemical splash goggles.

2. Measure the temperature of each of these solutions, using the same thermometer. However, rinse the thermometer and dry it after the first measurement. If the temperatures are not identical, cool the warmer solution by immersing the graduated cylinder in tap water, or warm the cooler solution with your hands. The temperatures should finally agree to within ±0.2°C. Record the mean temperature. This is the initial temperature, t_i.
3. Add the acid to the calorimeter. In order to account for incomplete draining, record the volume of the solution that remains in the graduated cylinder.
4. Add the base to the calorimeter. You will read the volume of any solution that remains in the cylinder in a subsequent step.
5. Immediately place the top on the calorimeter and begin stirring.
6. Record the temperature to the nearest 0.1°C after 30 s and every 30 s thereafter for 4 min.
7. During this time, measure and record the volume of the base remaining in the graduated cylinder from Step 4.
8. Plot the temperature of the calorimeter against the time the measurement was taken, using one of the pieces of graph paper that are available.

   Use a straight line to extrapolate your results to the time of mixing (time = 0 s). Record the extrapolated temperature. This is the final temperature, t_f.

9. Calculate q(system) using 4.184 J/(g • °C) and 1.0 g/mL for the specific heat and density of the solution, respectively, and 1.0×10^1 J/°C for the heat capacity of the calorimeter.
10. Calculate the enthalpy change, ΔH, from q(system) and the number of moles of the acid or base. If you used unequal volumes of the acid and base solutions, use the number of moles of the limiting reactant.
11. If time permits, repeat the measurement using the same acid and base and the same piece of graph paper to obtain t_f. Obtain the mean value of ΔH.

12. Repeat Steps 1 through 11 for the NH_4Cl–NaOH acid–base pair. Use a new piece of graph paper to graph your results and obtain t_f for this reaction.

Date ______________________ Student Name ______________________
Course/Section ______________________ Team Members ______________________
Instructor ______________________ ______________________

Thermochemistry and Hess's Law

Prelaboratory Assignment

1. Give chemical equations for the reactions that will occur during this experiment.

2. Define the system and the surroundings for these reactions. If water is one of the products, does it belong in the system or in the surroundings?

3. a. When solutions of two reactants were mixed in a coffee-cup calorimeter, the following temperatures were recorded as a function of time. Plot the data on one of the available pieces of graph paper. Obtain t_f, the final temperature, by extrapolating to the time of mixing (time = 0 s) with a straight line through the data to the point where it intersects the y-axis of your graph. The initial temperature, t_i, was 24.3°C.

Time (s)	*t* (°C)	Time (s)	*t* (°C)
30	40.4	150	40.5
60	40.8	180	40.4
90	40.7	210	40.3
120	40.6	240	40.2

t_f :__________

b. Is this an exothermic or an endothermic reaction? Why?

c. Why does the temperature increase, reach a maximum, and then decrease?

4. What special safety precautions must be observed during this experiment?

Date	______________	Student Name	______________
Course/Section	______________	Team Members	______________
Instructor	______________		______________

Thermochemistry and Hess's Law

Results

1. *Enthalpy change (NaOH–HCl or NH_3–HCl)*

Concentration of HCl: ______________

Concentration and identity of base: ______________

____

Trial	1	2
	______________	______________
Volume of acid in cylinder:		
Before pouring (mL)	______________	______________
After pouring (mL)	______________	______________
Volume added (mL)	______________	______________
Volume of base in cylinder:		
Before pouring (mL)	______________	______________
After pouring (mL)	______________	______________
Volume added (mL)	______________	______________
t_i (°C)	______________	______________

continued...

Trial	1	2
	________	________
Temperature (°C) after		
30 s	________	________
60 s	________	________
90 s	________	________
120 s	________	________
150 s	________	________
180 s	________	________
210 s	________	________
240 s	________	________
t_f(°C)	________	________
q(system) (J)	________	________
Moles of limiting reactant	________	________
ΔH (kJ/mol)	________	________
Mean ΔH (kJ/mol)	________	

Calculations:

Student name: ________________ Course/Section: __________________ Date: ________________

2. *Enthalpy change ($NaOH$–NH_4Cl)*

Concentration of NH_4Cl: __________________

Concentration of NaOH: __________________

Trial	**1**	**2**
Volume of acid in cylinder:		
Before pouring (mL)	__________________	__________________
After pouring (mL)	__________________	__________________
Volume added (mL)	__________________	__________________
Volume of base in cylinder:		
Before pouring (mL)	__________________	__________________
After pouring (mL)	__________________	__________________

Trial	1	2
Volume added (mL)	________	________
t_i (°C)	________	________
Temperature (°C) after		
30 s	________	________
60 s	________	________
90 s	________	________
120 s	________	________
150 s	________	________
180 s	________	________
210 s	________	________
240 s	________	________
t_f(°C)	________	________
q(system) (J)	________	________
Moles of limiting reactant	________	________
ΔH (kJ/mol)	________	________
Mean ΔH (kJ/mol)	________	

Calculations:

Student name: ________________ Course/Section: ________________ Date: ________________

3. *Pooled results* (Include your own.)

Reaction			**ΔH (kJ/mol)**			**Mean ΔH**
NaOH–HCl	______	______	______	______	______	
	______	______	______	______	______	
	______	______	______	______	______	______
NH_3–HCl	______	______	______	______	______	
	______	______	______	______	______	
	______	______	______	______	______	______
NaOH– NH_4Cl	______	______	______	______	______	
	______	______	______	______	______	
	______	______	______	______	______	
	______	______	______	______	______	
	______	______	______	______	______	______

Questions

1. a. Use Hess's law and the measured mean enthalpy changes for the NaOH–HCl and NH_3–HCl reactions to calculate the enthalpy change to be expected for the reaction

$$NaOH + NH_4Cl \rightarrow NaCl + NH_3 + H_2O$$

   b. Compare your experimental value with the one you have just calculated. The correct value is only –3.9 kJ/mol. Try to explain any discrepancy between the experimental and calculated values and between these values and the correct value.

2. Calculate the enthalpy change to be expected for the reaction

$$NaCl(s) \rightarrow NaCl(aq)$$

where (*s*) and (*aq*) mean solid and aqueous, respectively. Use Hess's law, one of the three enthalpy changes that was measured in this experiment, and the data from the following table.

Reaction*	**ΔH (kJ/mol)**
$1/2H_2(g) + 1/2Cl_2(g) \rightarrow HCl(g)$	–92.3
$Na(s) + 1/2O_2(g) + 1/2H_2(g) \rightarrow NaOH(s)$	–426.8
$Na(s) + 1/2Cl_2(g) \rightarrow NaCl(s)$	–411.1
$H_2(g) + 1/2O_2(g) \rightarrow H_2O(l)$	–285.8
$HCl(g) \rightarrow HCl(aq)$	–75.2
$NaOH(s) \rightarrow NaOH(aq)$	–41.8

* (*g*) = gas, (*l*) = liquid, (*s*) = solid, and (*aq*) = aqueous.

Calculations:

Student name: ________________ Course/Section: ________________ Date: ______________

3. Describe in detail an experiment using only hot and cold water that would enable you to verify that the heat capacity of your coffee-cup calorimeter is about 1.0×10^1 J/°C. Be specific about the procedure you would follow and the calculations you would need to perform.

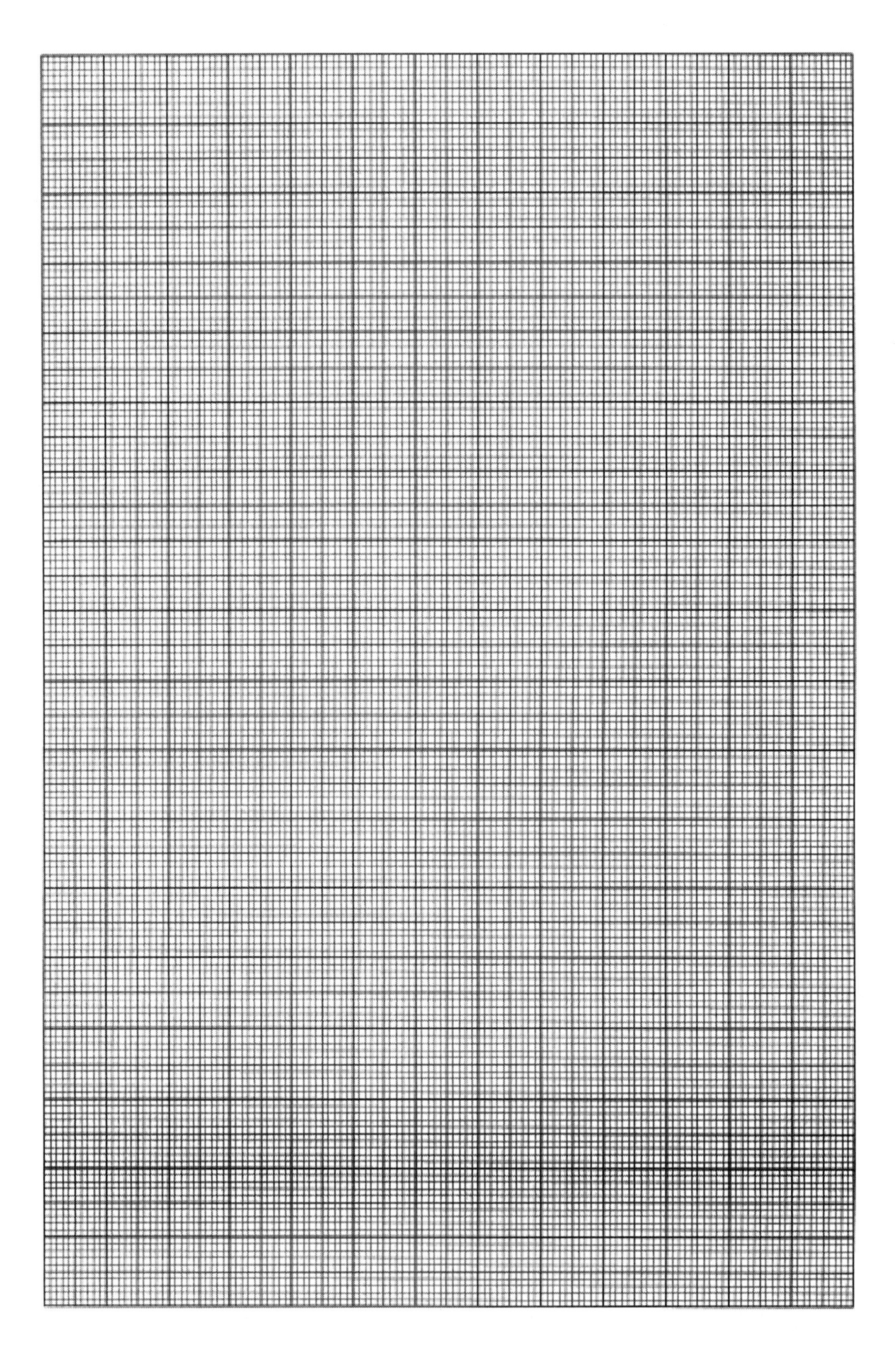

Student name: _______________ Course/Section: _______________ Date: _______________

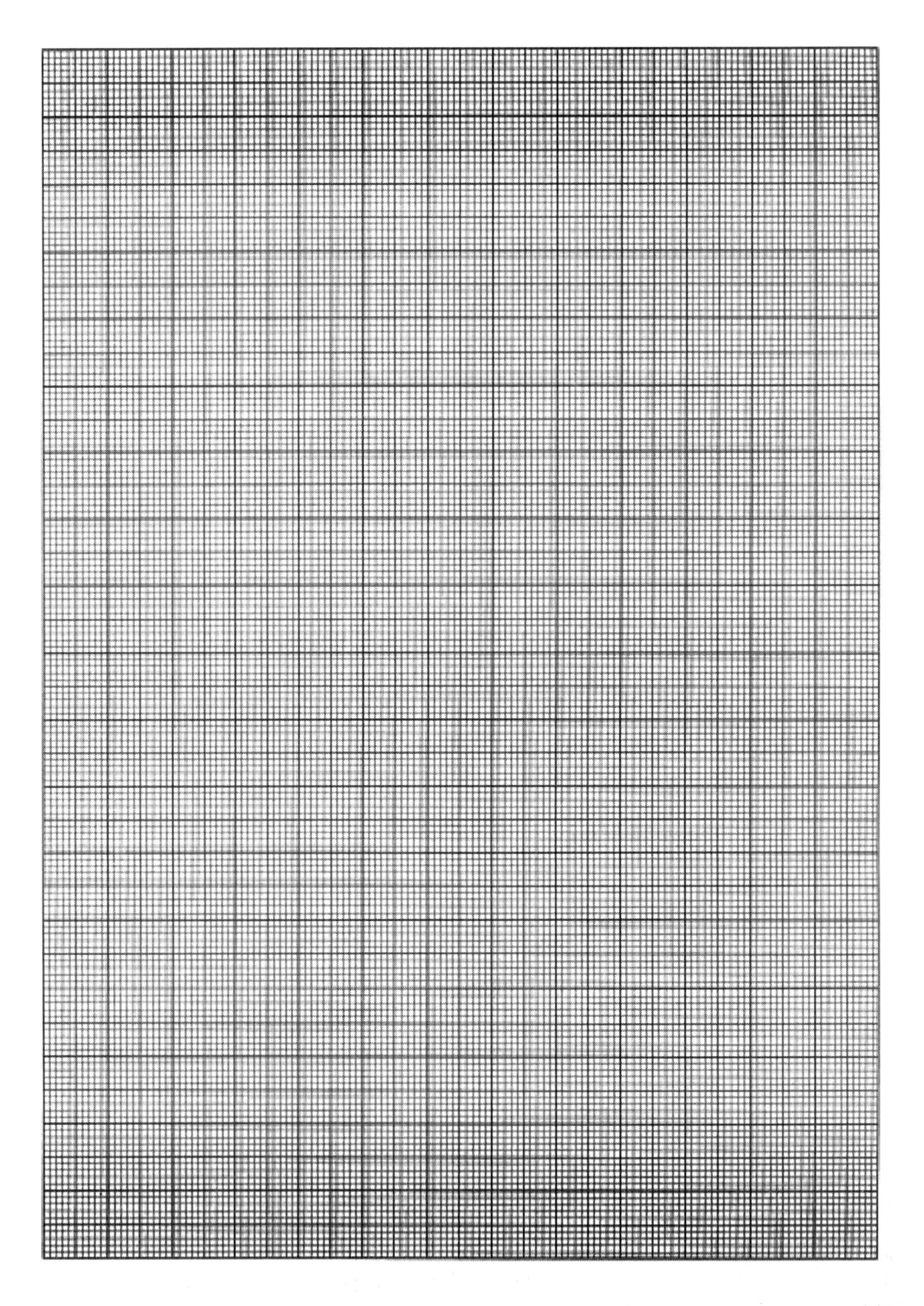

7. The Absorption Spectrum of Cobalt(II) Chloride

Introduction

When hydrogen and other gaseous substances are heated, they will *emit* light with a few characteristic frequencies (Ebbing/Gammon, Section 7.3). In contrast, matter will also *absorb* discrete frequencies of light. This process is exactly the opposite of emission.

During the absorption of light, an electron undergoes a transition from a lower-energy (usually the lowest-energy) level to a higher-energy level. The electron gains energy in this process by absorbing a photon, whose energy (E) is given by Einstein's equation, $E = h\nu$. The energy of the photon corresponds to the difference in energy between the higher-energy and lower-energy levels. As a result of the transition, a component of light with a frequency ν is absorbed, and other frequencies are transmitted.

Some further aspects of the absorption of light by solutions of colored substances are discussed in Appendix C. The amount of light absorbed by a substance in solution is dictated by the concentration of that substance; absorption decreases as the concentration is decreased by dilution. This statement is expressed mathematically in Beer's law (see Appendix C). You must understand the calculation of the concentration of a solution after dilution (Ebbing/Gammon, Section 4.8) before you can use or appreciate this law.

Purpose

You will investigate the absorption of light by a series of solutions of cobalt(II) chloride ($CoCl_2$). You will also receive a solution of this substance whose concentration is unknown but too large for an accurate measurement of the absorption of light. Your task will be to determine the concentration of this solution by diluting the solution, measuring the absorbance of the diluted solution and then calculating the concentration in the original solution.

Concept of the Experiment

The absorbance of a 0.150 *M* solution of $CoCl_2$ will be measured before and after a series of dilutions. For each dilution you will calculate the new concentration of $CoCl_2$. According to Beer's law, the slope of the best straight line from a graph in which absorbance is plotted against the concentration of $CoCl_2$ will give you k (the constant from Beer's law) for $CoCl_2$ under your experimental conditions. You will obtain the slope by a method called *linear regression*. This method is discussed in Appendix C. You may do this calculation by hand, or, if your laboratory instructor wishes, you may use a computer and the Internet.

You will also receive a solution of $CoCl_2$ whose concentration is unknown but so large that the absorbance is too great to be reliably measured. You must determine the concentration of $CoCl_2$ in this solution from a measurement of a diluted solution. It will be up to you to discover an appropriate dilution.

Before you measure any of these absorbances, however, you will need to find the correct wavelength of light to use for the measurements. This wavelength, which will be the wavelength in the absorption spectrum of $CoCl_2$ at which the maximum absorbance occurs, will give you the maximum sensitivity for each measurement.

Procedure

Getting Started

1. Your laboratory instructor may ask that you work in groups rather than alone.
2. Obtain directions from your laboratory instructor for discarding the solutions that you will use in this experiment.
3. Obtain instructions for using your spectrophotometer.
4. Obtain your unknown.

Making the Measurements

1. Mark each of 7 dry 18 × 150-mm test tubes with one of a series of identification numbers running from 1 to 7.
2. Use Mohr pipets to make the additions of 0.150 *M* $CoCl_2$ and distilled water shown in the table below.

CAUTION: Never use your mouth to draw liquid into a pipet, even if the liquid is water. Use a rubber suction bulb or some other suction device.

Test Tube No.	0.150 *M* $CoCl_2$ Solution (mL)	Distilled Water (mL)
1	5.0	0
2	4.0	1.0
3	3.5	1.5
4	3.0	2.0
5	2.5	2.5
6	2.0	3.0
7	1.0	4.0

3. Thoroughly mix the contents of each test tube by gently shaking the test tube from side to side. Do not use one of your fingers as a stopper.
4. Obtain the absorption spectrum of aqueous $CoCl_2$ by using the contents of the first test tube and measuring the absorbance at intervals of 25 nm between 400 and 600 nm. Do not dispose of this solution. Record your results.
5. From these measurements, select the wavelength at which the absorbance is largest. This wavelength will provide maximum sensitivity. Use this wavelength for all subsequent measurements.
6. Remeasure the absorbance of the contents of the first test tube at this wavelength. Measure the absorbances of the contents of the remaining test tubes. Record your results.
7. Label a piece of the available graph paper so that absorbance (A) appears on the vertical axis and the concentration (c) of $CoCl_2$, in mol/L, appears on the horizontal axis. Use Figure C.2 in Appendix C as an exact model.

8. Enter each point on the graph as a small, blackened circle. Do the data appear to conform to a straight line? Use linear regression (Appendix C) to calculate the slope of the best straight line that satisfies these points. If you wish, draw a straight line with this slope on your graph. This line should pass through the origin ($A = 0$, $c = 0$).

9. Now you can begin to work on your unknown. Use the following guidelines in establishing the molar concentration of $CoCl_2$ in this solution.

   a. A few milliliters of this solution must be diluted until the absorbance lies within the range of absorbances that you found in Step 6.

   b. To establish the correct dilution, measure the absorbance after each of a series of successive dilutions. Your final result here may have a large experimental error, because errors will accrue during several dilutions.

   c. To eliminate the accumulation of experimental errors, prepare a new sample and obtain the desired absorbance in one dilution rather than a series of dilutions. Clearly, this dilution must be equivalent to the overall dilution obtained in the previous set of trials.

10. Calculate the concentration of $CoCl_2$ in the diluted solution, using the absorbance and k. Calculate the concentration of $CoCl_2$ in the original unknown.

Date ____________________ Student Name ____________________
Course/Section ____________________ Team Members ____________________
Instructor ____________________ ____________________

The Absorption Spectrum of Cobalt(II) Chloride

Prelaboratory Assignment

1. a. What is Beer's law?

   b. According to this law, what will be the value of the slope if absorbance is plotted against concentration?

2. The following data were collected when the absorbances of a series of solutions containing $NiCl_2$ were measured:

Test Tube No.	1	2	3	4
Volume of 0.390 M $NiCl_2$ solution (mL)	4.0	3.0	2.0	1.0
Volume of H_2O (mL)	0	1.0	2.0	3.0
Concentration of $NiCl_2$ (M)	______	______	______	______
Absorbance (A)	0.858	0.644	0.429	0.215

   a. Complete the table by calculating the concentration of $NiCl_2$ in each solution. Show your calculations below.

b. Prepare a graph in which the absorbances given in the table are plotted against the concentration of $NiCl_2$ that you have just calculated. Follow Steps 7 and 8 under Making the Measurements. Use the space below to calculate k by hand using linear regression. Repeat the calculation using the tool available online at the student website if your laboratory instructor wishes.

c. The absorbance of a solution of $NiCl_2$ of unknown concentration is found to be 0.388. Using Beer's Law and the value of 'k" determined in part b, determine the concentration of this solution.

3. What safety rule must be observed during this experiment?

Date	____________	Student Name	____________
Course/Section	____________	Team Members	____________
Instructor	____________		____________

The Absorption Spectrum of Cobalt(II) Chloride

Results

1. *The absorption spectrum of aqueous $CoCl_2$*

Wavelength (nm)	*A*	Wavelength (nm)	*A*	Wavelength (nm)	*A*
400	________	475	________	550	________
425	________	500	________	575	________
450	________	525	________	600	________

The maximum absorbance occurs at ___________ nm.

2. *The absorbance as a function of the concentration (c)*

Wavelength to be used for measurements: ____________ nm

Test Tube No.	A	c
1	____________	____________
2	____________	____________
3	____________	____________
4	____________	____________
5	____________	____________
6	____________	____________
7	____________	____________

Calculations of c:

Student name: ________________ Course/Section: __________________ Date: _______________

3. *Calculations for the slope* (Include your graph.)

k = ______________________ (Give units.)

4. *Identifying the unknown concentration*

Unknown no.: _____________________

a. Trial solutions

mL of Unknown	mL of H_2O	*A*
____________	____________	____________
____________	____________	____________
____________	____________	____________
____________	____________	____________

b. Final determination

Volume of unknown (mL): ____________

Volume of H_2O (mL): ____________

A = ____________

c = ____________ M

Undiluted solution: c = ____________ M

Calculations:

Question

1. a. Write a brief procedure outlining how would you prepare 250 mL of a 0.150 M solution of $CoCl_2$ from solid $CoCl_2$ and distilled water.

Student name: ________________ Course/Section: ________________ Date: ________________

b. Using the solution from part a and distilled water, how would you prepare 100 mL of a 0.0600 *M* solution of $CoCl_2$?

c. What absorbance would you expect from the solution from part b at the wavelength of maximum sensitivity?

Student name: ________________ Course/Section: __________________ Date: ________________

8. Solubilities Within a Family

Introduction

The periodic table is arranged in such a way that the electron configurations of the elements display a periodic variation (Ebbing/Gammon, Section 8.6). The same kind of outer configuration occurs within a group (vertical column), period after period (horizontal row). For example, the outer-electron configuration of an alkaline earth metal (Group 2A) is always ns_2, no matter what period the element occupies.

The periodicity in the outer-electron configurations is responsible for the periodic law: When the elements are arranged by atomic number, their physical and chemical properties vary periodically. Because of this periodicity, elements within the same group form compounds that have the same general formula. Thus all the alkaline earth metals form oxides with the same general formula.

The periodic law, however, does not imply that all the properties of the elements within a group will be identical. Trends in properties are usually found instead. Thus atomic size increases smoothly going down a column of main-group elements, whereas ionization energy decreases (Ebbing/Gammon Section 8.6).

Purpose

This experiment will give you the opportunity to look for trends in the relative solubilities of some compounds of the alkaline earth metals. You will also compare the solubilities of these compounds with those of some similar compounds of lead, a metal in Group 4A.

Concept of the Experiment

The solubility of a compound in a liquid is the maximum amount of that compound that will dissolve at a fixed temperature. In this experiment you will be examining the *qualitative aspects* of solubility. Qualitative aspects are properties of a compound that you can observe but don't measure. A For example, the blue color is a qualitative aspect of cobalt chloride while a molar mass of 129.839 grams/mol is a quantitative aspect of the same compound. You will see that the qualitative terms *soluble* and *insoluble* can be used to describe a compound's observed solubility.

There are two ways to determine the solubility of a compound qualitatively. First, we can take the compound directly from a bottle, place the compound in the desired liquid, and observe the solubility. If the solid compound dissolves in the desired liquid, we say that the compound is "soluble" in the liquid. A second way to determine solubility is to synthesize (make) the compound by a chemical reaction in the liquid. If the compound precipitates from solution (appears as a solid), we know that it cannot be very soluble in the liquid. If it does not appear, it must be soluble. You will employ the second method in this experiment.

The alkaline earth metals and lead form nitrates, hydroxides, chlorides, bromides, and iodides with the general formulas $M(NO_3)_2$, $M(OH)_2$, MCl_2, MBr_2, and MI_2, respectively. You will use the reaction

$$M(NO_3)_2 + 2NaX \rightarrow MX_2 + 2NaNO_3$$

to determine the qualitative solubilities of the hydroxides, chlorides, bromides, and iodides in water. Each of these is represented by X in the general reaction. We know that $NaNO_3$ (sodium nitrate) must be very soluble because all nitrates are very soluble. Thus, if a precipitate appears, it can only be MX_2, and this must mean that this compound is not very soluble in water.

The alkaline earth metals and lead also form sulfates, carbonates, oxalates, and chromates with the general formulas MSO_4, MCO_3, MC_2O_4, and $MCrO_4$, respectively. These compounds will be prepared by the reaction

$$M(NO_3)_2 + Na_2Y \rightarrow MY + 2NaNO_3$$

where Y represents sulfate, carbonate, oxalate, and chromate. The solubilities of the MY compounds will be determined in exactly the same way as those of the MX compounds.

Because you will be examining only the qualitative aspects of the solubilities of these compounds, you will not observe trends directly. Instead, you will need to use deductive reasoning to conclude whether or not a trend exists. The following example should make this process easier.

Consider the solubilities of the chromates shown in Table 8.1. Let us pretend for the moment that we do not know these solubilities. What would you find if you were to use a chemical reaction to produce 1.0
$\times$ 10_{-4} mol of each substance in 1000 mL of water? Note that only the solubility of $BaCrO_4$ has been exceeded. As a result, all of the $MgCrO_4$, $CaCrO_4$, and $SrCrO_4$ would remain in solution, and a precipitate of $BaCrO_4$ would appear. This experiment would force us to admit the possibility that if a trend exists, the solubilities may decrease as the atomic number of the Group 2A metal increases. As Table 8.1 shows, this is correct.

Table 8.1

Solubilities of Some Chromates (mol/1000 mL H_2O)

$MgCrO_4$	$CaCrO_4$	$SrCrO_4$	$BaCrO_4$
9.9	1.2	5.9×10^{-3}	1.1×10^{-5}

This example is not intended to portray a general trend. In some instances, solubilities will increase as the atomic number of the Group 2A metal increases.

Procedure

Getting Started

1. Obtain 5 small test tubes.
2. Use a 5-mL or a 10-mL graduated cylinder to place 1 mL of distilled water in each of these test tubes. Mark the height of the water in each test tube with a small piece of tape or a marking pencil. Add an additional 1 mL to each test tube, and mark the new height of the water in each. Pour the water into a sink.
3. Obtain directions from your laboratory instructor for discarding the solutions that you will use in this experiment.
4. Observe the following safety precaution during the remainder of this experiment.

CAUTION: Wash your hands thoroughly after using solutions containing lead, barium, or oxalate because they are poisonous.

Determining the Qualitative Solubilities

1. Mark the test tubes with identification numbers (1 through 5).

2. Using the lower marks on the test tubes as guides, add 1 mL of 0.1 *M* $Mg(NO_3)_2$ to the first test tube, 1 mL of 0.1 *M* $Ca(NO_3)_2$ to the second, 1 mL of 0.1 *M* $Sr(NO_3)_2$ to the third, 1 mL of 0.1 *M* $Ba(NO_3)_2$ to the fourth, and 1 mL of 0.1 *M* $Pb(NO_3)_2$ to the fifth. Observe the color and clarity of each solution.

3. Using the upper marks as guides, add 1 mL of 1 *M* NaOH to each test tube. Shake each tube gently and wait about 30 s. Note the colors of all precipitates. If any of the precipitates are virtually colorless, they may be difficult to see. Be observant! A precipitate may appear as a change in the clarity of the solution. Note the relative amounts of the precipitates. If the solubility of a compound is barely exceeded, only a trace of a precipitate will appear and the solution may appear "cloudy". Record your observations.

   CAUTION: Do not use your finger as a stopper.

4. Discard the contents of each of the test tubes as directed by your laboratory instructor. Wash the test tubes carefully, and rinse them with distilled water.

5. Repeat Steps 1 through 4 six more times, in turn adding (in step 3) 1 *M* NaCl, 0.1 *M* NaBr, 0.1 *M* NaI, 0.1 *M* Na_2SO_4, 0.1 *M* Na_2CO_3, and 0.1 *M* $Na_2C_2O_4$ instead of the solution of NaOH.

Date ______________________ Student Name ______________________
Course/Section ______________________ Team Members ______________________
Instructor ______________________ ______________________

Solubilities Within a Family

Prelaboratory Assignment

1. Give the names and symbols of the alkaline earth metals that you will encounter in this experiment.

2. a. What is the general electron configuration of the alkaline earth metals?

   b. What is the electron configuration of lead?

   c. How do the formulas of the oxides formed by the alkaline earth metals compare with those of lead.

3. Give general formulas for the following compounds with alkaline earth metals or lead:

   a. A sulfate

   b. A chloride

   c. A carbonate

   d. An iodide

   e. A bromide

   f. An oxalate

   g. A hydroxide

   h. A nitrate

   i. A chromate

4. What general method will be used to examine the qualitative solubility of the alkaline earth metal and lead compounds in this experiment?

5. What safety precautions must be observed in this experiment?

Date ______________________ Student Name ______________________
Course/Section ______________________ Team Members ______________________
Instructor ______________________ ______________________

Solubilities Within a Family

Results

	$Mg(NO_3)_2$	$Ca(NO_3)_2$	$Sr(NO_3)_2$	$Ba(NO_3)_2$	$Pb(NO_3)_2$
NaOH					
NaCl					
NaBr					
NaI					
Na_2SO_4					
Na_2CO_3					
$Na_2C_2O_4$					

Questions

1. Based on your data and using deductive reasoning, what can you conclude about the trends in the solubilities of the compounds of the alkaline earth metals?

2. a. Compare the solubilities of the lead compounds with those of the alkaline earth metals. How are the solubilities similar, and how do they differ?

   b. As you have seen, lead can form the same kinds of compounds as the alkaline earth metals. Nevertheless, the solubilities may differ markedly. What do you think is the reason for these differences?

9A. The Identity of an Insoluble Precipitate

Introduction

The properties of any substance depend in part on the chemical bonds that hold the atoms of the substance together. The consequences of this dependence are very important in chemical reactions. Because bonds are formed or broken during a reaction, the properties of product molecules differ from those of reactant molecules. If there is a significant difference in these properties, a distinct signal that a chemical reaction has occurred can usually be observed.

One easily observed signal of a chemical reaction is the formation of an insoluble precipitate. This experiment deals with a *quantitative* interpretation of a reaction in which this signal has appeared.

Purpose

In this experiment, you will examine the reaction between $Ba(NO_3)_2$ and NH_2SO_3H (sulfamic acid) in a hot solution. The identity of the insoluble substance that results from the reaction will be determined the mass of the substance with the masses of the possible products of the reaction. the.

Concept of the Experiment

Known quantities of $Ba(NO_3)_2$ and NH_2SO_3H will be allowed to react in boiling water. Certain covalent bonds in the molecules of sulfamic acid will slowly break , and a polyatomic anion will be formed. This anion will combine with the Ba^{2+} cations from $Ba(NO_3)_2$ to form an ionic substance that appears as a white precipitate.

All of this precipitate must be separated by gravity filtration because its mass must be measured. Make sure that you read about this type of filtration in the Introduction section of this manual. You will also have an opportunity to practice this type of filtration during the experiment. Take full advantage of this opportunity!

You will need to accurately know the mass of the precipitate to determine its identity therefore you must remove all of the residual water by drying it completely. You will know when the precipitate is dry because its mass will reach a constant value within acceptable error (±0.002 g, or any other range specified by your laboratory instructor).

The precipitate will be one of three possible compounds:

1. $Ba(NH_2SO_3)_2$ (barium sulfamate)
2. $BaSO_4$ (barium sulfate)
3. $Ba(NH_2)_2$ (barium amide)

You will determine which one it is from the masses of the precipitate and the limiting reactant (Ebbing/Gammon, Section 3.8).

Because of time constraints, you will be unable to do the experiment more than once. Be careful; avoid major mistakes and systematic errors. Relatively small random errors can be tolerated, however, because you will pool your results with those obtained by other students. As a result, you will have enough data to determine the precision by the method shown in Appendix A.

Procedure

Getting Started

1. Make sure that you have read about gravity filtration in the Introduction section of this manual.
2. If you do not understand how to fold the filter paper, ask your laboratory instructor for help.
3. Obtain directions for discarding the solution that you will use in this experiment from your laboratory instructor.

Initiating the Reaction

1. You will need about 1.3–1.4 g of $Ba(NO_3)_2$ and about 2.4–2.5 g of NH_2SO_3H. One of these substances is the limiting reactant, as you showed in the Prelaboratory Assignment. You should measure its mass with your most precise balance. Your laboratory instructor will give you instructions for using the balance. Use weighing paper in all cases. Record each mass that you measure.
2. Transfer both samples to a 250-mL beaker and add 150 mL of distilled water using a graduated cylinder.

   CAUTION: Wash your hands thoroughly after using the solution containing barium, because it is poisonous.

3. Using the rubber end of a glass stirring rod equipped with a rubber policeman, stir the mixture until most of the solids have dissolved. The remainder will dissolve when the solution is heated. Do not remove the stirring rod from the beaker.
4. Mark the volume of the solution in the beaker with a marking pencil so that a constant volume can be maintained throughout the course of the reaction.
5. Set up a ring stand with an iron ring, and put a piece of wire gauze on the ring. Adjust the height of the ring so that the wire gauze will be in the hottest part of the flame from a laboratory burner. Do not light the burner until you have made this adjustment.

   CAUTION: Avoid burning your fingers. Do not touch the iron ring or the wire gauze at any time while the solution is being heated.

6. Place the beaker on the wire gauze, and heat the solution to a *gentle* boil. Allow the solution to boil for about 40 min after the first cloudiness appears.
7. Stir the solution occasionally without removing the stirring rod. Add small amounts of distilled water as needed to maintain the original volume of liquid in the beaker.
8. While the solution is boiling, practice filtration.

Practicing Filtration

1. Obtain about 0.5 g of powdered $CaCO_3$ (calcium carbonate) on a piece of weighing paper, using the balance provided in the laboratory.
2. Transfer the sample of $CaCO_3$ to a large beaker. Add 150 mL of distilled water. Stir the mixture thoroughly with the rubber end of another glass stirring rod equipped with a rubber policeman.

3. Filter the mixture using gravity filtration, a technique described in the Introduction section of this manual.
   No trace of $CaCO_3$ should remain in the beaker after filtration is complete.

4. If you wish, repeat steps 1 through 3 with another sample of $CaCO_3$ in water.

Finishing the Experiment

1. When the period of heating is completed, turn off the laboratory burner and cool the beaker to room temperature using cold water or ice.

2. While the beaker and its contents are cooling, obtain and record the mass of a piece of filter paper using your most precise balance. You will use this filter paper to separate the precipitate from the solution.

3. Filter the cooled mixture containing the precipitate, using the technique that you have practiced.

4. Use a metal spatula to loosen the edge of the filter paper from the filter funnel.

5. Carefully transfer the filter paper and its contents to a beaker that has been labeled with your name. The paper should be upright and never upside down.

6. Place the beaker in an oven at 85°C and allow it to remain there for at least 1 h.

7. During this time, you can review your conclusions about the limiting reactant (see Prelaboratory Assignment). You can also calculate the theoretical yields (Ebbing/Gammon, Section 3.8) of $Ba(NH_2SO_3)_2$, $BaSO_4$, and $Ba(NH_2)_2$ (see Question 2a).

8. After 1 h, remove the beaker from the oven, using tongs, and allow it to cool. Obtain and record the mass of the filter paper and its contents, using your most precise balance.

9. Return the paper to the beaker and the beaker to the oven for an additional 5 min. Remove the beaker, cool the paper, and obtain its mass once again. Continue until a constant mass (±0.002 g, or any other range stipulated by your laboratory instructor) is reached.

10. Calculate the value of the ratio

    Mass of precipitate/mass of limiting reactant

    and share this information with other students in the class as directed by your laboratory instructor. Record their ratios and compare them with your own.

CAUTION: Before you leave the laboratory, make sure that your gas outlet and those of your neighbors are closed.

9B. From Oil of Wintergreen to Salicylic Acid

Introduction

What happens during a chemical reaction? Reactant molecules are converted to product molecules as chemical bonds are formed or broken. The properties of a molecule depend in part on the bonds within the molecule, as a result, the properties of the reactants differ from those of the products. These differences are often large.

This experiment deals with a chemical reaction in which there is a significant and easily observed difference between the properties of a reactant and those of a product: The reactant is a liquid and the product is a solid.

Some of the substances in this experiment are called organic compounds. *Organic chemistry* deals with compounds in which carbon is the principal element.

Purpose

You will examine the chemical reaction in which methyl salicylate (oil of wintergreen), a liquid, is converted to salicylic acid, a solid. The actual yield of this product will be compared with the theoretical yield.

Some Interesting Facts

Methyl salicylate, well known for its characteristic wintergreen odor, is used as a flavoring and as an ingredient in various liniments. Salicylic acid, which is used in the preparation of aspirin, shares aspirin's ability to relieve pain.

Concept of the Experiment

When methyl salicylate ($HOC_6H_4COOCH_3$) is heated in water, salicylic acid (HOC_6H_4COOH) and methyl alcohol (CH_3OH) are formed slowly. This reaction is shown in Figure 9B.1. You can obtain a faster reaction if you replace the water with an aqueous solution of sodium hydroxide. When the reaction is complete, the sodium hydroxide can be removed chemically by the addition of sulfuric acid. Although the conversion of methyl salicylate to salicylic acid becomes slightly more complicated under these conditions, the overall results are virtually identical to those shown in Figure 9B.1.

FIGURE 9B.1

The reaction of methyl salicylate (oil of wintergreen) with water to give salicylic acid and methyl alcohol.

Salicylic acid is not very soluble in an aqueous solution of sulfuric acid. As a result, salicylic acid precipitates when you add sulfuric acid. You will use suction filtration to separate the precipitated solid from the solution. This filtration technique is described in the Introduction section of this manual.

The product that you have recovered from this reaction may be crude or impure. This means it will contain products other than the salicylic acid. . You will use a technique called *recrystallization* to remove possible impurities. In this technique, a crude solid product from a reaction is dissolved in a liquid, and crystallization is induced. In the simplest use of the technique, purification will occur if the impurities are more soluble in the liquid than the desired compound is. The impurities will remain in solution after the purified product has crystallized.

After recrystallization, the crystals must be dried. There are two possible techniques. Your laboratory instructor will tell you which one to use based upon the time available.

You will also have an opportunity to calculate the theoretical yield and the percentage yield of salicylic acid (Ebbing/Gammon, Section 3.8) on the basis of the mass of the limiting reactant. You will probably obtain a percentage yield greater than 80%.

Procedure

Getting Started

1. Make sure that you have read and understood the description of suction filtration in the Introduction section of this manual.
2. Observe the following safety precautions:

   CAUTION: The solutions of sodium hydroxide and sulfuric acid that you will encounter must be handled carefully. These solutions can cause chemical burns, in addition to ruining your clothing. If you spill one of these solutions on you, wash the contaminated area thoroughly and report the incident to your laboratory instructor. You may require further treatment.

3. Add 60 mL of distilled water to a 250-mL beaker using a graduated cylinder. Mark the location of the upper surface of the water on the outer wall of the beaker, using a marking pencil. You will use this beaker for the reaction. First, however, discard the water and dry the beaker.
4. Ask your laboratory instructor which technique you should use to dry the crystals of salicylic acid.

Doing the Reaction

1. Obtain 4.0 mL of methyl salicylate in a clean, dry 10-mL graduated cylinder. Measure and record the mass, using the laboratory balance. Your laboratory instructor will provide you with instructions for using these balances.
2. Add as much of the methyl salicylate from Step 1 as you can to the dry, marked 250-mL beaker.
3. Measure and record the mass of the graduated cylinder. Calculate the mass of the methyl salicylate that you poured into the beaker.
4. Add 40 mL of 6 *M* NaOH from a clean graduated cylinder. A white precipitate will form immediately. This precipitate is not the desired product. The ionic compound in this precipitate is $NaOC_6H_4COOCH_3$.
5. Stir the mixture thoroughly with a glass stirring rod. Do not remove the stirring rod.
6. Set up a ring stand with an iron ring, and place a piece of wire gauze on the ring. Adjust the height of the ring so that the wire gauze will be in the hottest part of the flame from a laboratory burner. Do not light the burner until you have made this adjustment.

CAUTION: Avoid burning your fingers. Do not touch the iron ring or the wire gauze at any time while the contents of the beaker are being heated.

7. Place the beaker on the wire gauze and secure the beaker to the ring stand with either a large clamp or a larger ring. Heat the mixture to a *gentle* boil. Stir the solution occasionally. The precipitate will dissolve.
8. Continue to boil the solution for 15 min after the precipitate has dissolved. Use a stream of distilled water from a plastic wash bottle to rinse any solids from the inner walls of the beaker into the solution. However, do not let the total volume in the beaker exceed 60 mL, as indicated by your mark.
9. Take this opportunity to assemble the glassware for suction filtration.
10. After the period of heating is completed, turn off the laboratory burner and cool the beaker in ice until it is only warm when you touch it.
11. Add 50 mL of distilled water to the beaker, and then *cautiously* add 50 mL of 8 M H_2SO_4 with stirring. A white precipitate of the crude product should form during the addition.
12. Cool the beaker again in ice until it is cold. At the same time, cool 50 mL of distilled water in an Erlenmeyer flask. You will use the cold water to wash the precipitate.
13. Filter the cold mixture by suction filtration, and wash the crude product with the cold water.
14. The suction should be continued for several minutes to partially dry the precipitate.
15. While you are drying the product in the filter, wash and dry the 250-mL beaker.

Recrystallizing the Product

1. Tilt the Büchner funnel over the 250-mL beaker. Use a metal spatula to carefully separate the precipitate from the filter paper and to transfer the precipitate to the beaker.
2. Add 100 mL of distilled water to the beaker from a graduated cylinder.
3. Place the beaker on the wire gauze and secure the beaker to the ring stand with either a large clamp or a larger ring.. Heat the mixture until a gentle boil occurs. Stir with a stirring rod. Several minutes of boiling are usually required before all of the precipitate dissolves.
4. Turn off the laboratory burner, and allow the solution to cool for 5 min without disturbing it. You should see crystals of salicylic acid begin to appear. Record a description of their appearance.
5. Cool the beaker in ice until it is cold. At the same time, cool 50 mL of distilled water in an Erlenmeyer flask.
6. Filter the cold mixture by suction filtration. Wash the crystals with the cold water.
7. Continue the suction for 15 min.
8. While you are drying the product in the filter, wash and dry the 250-mL beaker during this time.

Drying the Crystals

1. Measure and record the mass of the dry 250-mL beaker, using the laboratory balance indicated by your instructor.
2. Transfer the crystals to the beaker, using the techniques that you used earlier.

3. If you have only a limited amount of time, follow Steps 4 through 7 and Steps 8 and 9. However, if it is possible to wait until the next laboratory period, follow Steps 8 through 10. Your laboratory instructor will let you know which trying method you are to use for this experiment.
4. Cover the beaker with a clean watch glass. Place the beaker on the wire gauze.
5. Follow the directions in this step carefully. Brush the flame back and forth across the underside of the wire gauze for about 30 s. Do not allow the flame to linger in any one place. You should see a very thin layer of water condense on the watch glass.
6. Allow the beaker to cool briefly, and remove the watch glass. Measure and record the mass of the beaker and its contents.
7. Replace the beaker and watch glass on the wire gauze. Repeat Steps 4, 5, and 6. Repeat until the mass does not change significantly.
8. Write your name on the beaker with a marking pencil. Cover it with a clean watch glass, and set it aside so that the crystals can dry in the air.
9. When the crystals are dry, remove the watch glass. Measure and record the mass of the beaker and its contents.

CAUTION: Do not ingest your sample under any circumstances.

10. Show your sample to your laboratory instructor. Obtain instructions for its disposal.

CAUTION: Before you leave the laboratory, make sure that your gas outlet and those of your neighbors are closed.

Date ____________________ Student Name ____________________
Course/Section ____________________ Team Members ____________________
Instructor ____________________ ____________________

From Oil of Wintergreen to Salicylic Acid

Prelaboratory Assignment

1. Draw sketches of methyl salicylate and salicylic acid, showing every bond.

2. What is the limiting reactant in this experiment? Why? Assume water is the other reactant. The density of methyl salicylate is 1.18 g/mL at 20°C.

3. What safety precautions must be observed during this experiment?

Date ______________________ Student Name ______________________
Course/Section ______________________ Team Members ______________________
Instructor ______________________ ______________________

From Oil of Wintergreen to Salicylic Acid

Results

Mass of methyl salicylate and graduated cylinder (g): ______________

Mass of graduated cylinder (g): ______________

Mass of methyl salicylate (g): ______________

Description of crystals of salicylic acid: ______________

Mass of salicylic acid and beaker (g): ______________

Mass of beaker (g): ______________

Mass of salicylic acid (g): ______________

Questions

1. a. What is the actual yield of salicylic acid?

   b. Calculate the theoretical yield. Remember that you deduced the limiting reactant in the Prelaboratory Assignment.

c. Calculate the percentage yield.

2. What covalent bonds in methyl salicylate and water were broken during the reaction? What bonds were formed?

10. Geometric Isomers

Introduction

The existence of geometric or *cis–trans* isomers is a consequence of the lack of rotation about double bonds. Maleic and fumaric acids, whose structures are shown in Figure 10.1, are a pair of geometric isomers.

Although their molecular formulas are identical ($C_4H_4O_4$), *cis* and *trans* isomers are entirely different compounds. They are different compounds because they have different geometries and because rotation about the π bond will not occur under ordinary conditions. You will note that maleic acid, the *cis* isomer, has a more congested array of atoms than does fumaric acid, the *trans* isomer. The crowding in the *cis* isomer occurs because both carboxylic acid groups (–COOH) are on the same side of the double bond. In contrast, these groups are on opposite sides of the double bond in the *trans* isomer.

Purpose

You will receive two unidentified samples. One will be maleic acid, and the other will be fumaric acid. The identities of these samples will become clear as you compare some of their properties.

FIGURE 10.1

The structure of (A) maleic acid, (B) fumaric acid, and (C) succinic acid.

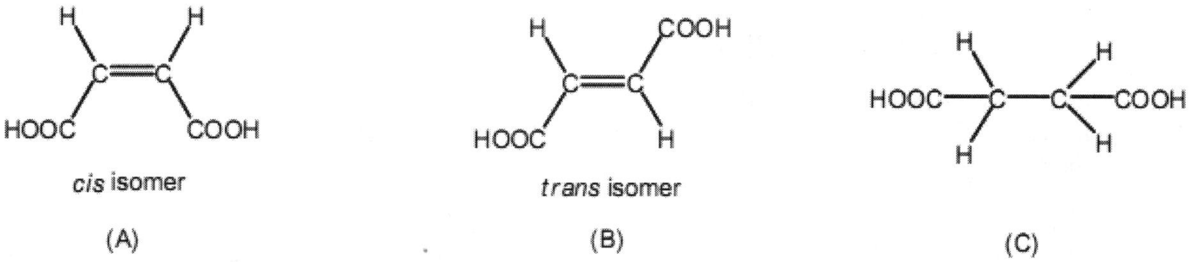

Isomerization

It is often possible to convert a crowded *cis* isomer to the *trans* isomer by a chemical reaction. This reaction requires an input of energy to rupture the π bond. Rotation about the remaining σ bond then occurs, and the less crowded arrangement can be obtained. After rotation, the π bond is formed once again. The name of this type of reaction is *isomerization.*

Concept of the Experiment

Maleic and fumaric acids are different compounds, so they will have entirely different properties. Their melting points and solubilities in water are compared in Table 10.1.

Table 10.1

Some Properties of Maleic and Fumaric Acids

Compound	Melting Point (°C)	Solubility in H_2O (g/1000 mL) at 25°C
Maleic acid	131	788
Fumaric acid	287	7

You will base the identification of your samples on a comparison of their melting points and their solubilities. You will not need to determine the absolute values of the melting points and solubilities of your samples because you are making a comparison between the two compounds. Instead, you will determine which sample has the lower melting point and which has the lower solubility in water. Your conclusions from this comparison should be confirmed when you determine which sample will undergo isomerization.

Procedure

Getting Started

1. Your laboratory instructor may ask you to work with a partner.
2. Make sure that you have read and understood the description of suction filtration in the Introduction section of this manual.
3. Obtain your unknown samples and 2 small test tubes.

Examining the Samples' Properties

1. Mark the 2 small test tubes so that you can recognize each sample. Place a pea-sized portion of one sample in the appropriate test tube and a similar-sized portion of the other sample in the remaining test tube.
2. Light a laboratory burner. Hold each test tube in a separate test tube holder with the same hand. Heat the samples simultaneously until one of them melts. Be sure that the tubes are not capped and that they do not point at anyone. Record which sample has melted. Allow the test tubes to cool before cleaning them. Discard the contents of each test tube as directed by your laboratory instructor.
3. Wash your test tubes, rinse them with distilled water, and dry them.
4. Repeat Step 1. Add 2 mL of distilled water to each test tube, and shake the tube gently for about 1 min. Record which sample has dissolved.
5. You should be able to identify each sample at this point by comparing your data with those shown in Table 10.1. The results of the next test should confirm the identities.

Attempting Isomerization

1. Obtain the mass of a piece of weighing paper. Your laboratory instructor will give you directions for using the balance. Add 1.0 g of the more soluble sample to this paper. Transfer this portion of the sample to a 250-mL Erlenmeyer flask.
2. Add 10 mL of distilled water to this flask and swirl gently to dissolve the sample.

3. Set up a ring stand with an iron ring, and put a piece of wire gauze on the ring. If possible, put this apparatus in a hood. If a hood is not available, use an inverted conical filter funnel connected by rubber tubing to a water aspirator to suck away HCl vapor, which will be generated in Steps 6 and 7. Secure the funnel to a ring stand with a clamp. Make sure that the rubber tubing is away from the flame of the laboratory burner.

4. Place a laboratory burner under the wire gauze, and adjust the height of the iron ring so that the wire gauze will be approximately in the hottest part of the flame. Do not light the burner until this adjustment has been made.

   CAUTION: Avoid burning your fingers. Do not touch the iron ring or the wire gauze at any time during heating. Be careful about touching the flask after heating; hot glass looks the same as cold glass. In case of burns, notify the lab instructor and cool the affected area. (Immerse in ice water if possible.)

5. Place the flask on the wire gauze, and heat the solution until the water boils.

6. Measure 10 mL of concentrated hydrochloric acid in a graduated cylinder.

   CAUTION: Concentrated acids must be handled carefully because they can cause severe chemical burns, in addition to ruining your clothes. Be sure your goggles are properly secured over your eyes. If you spill an acid on you, wash the contaminated area thoroughly with tap water and report the incident to your laboratory instructor. You may require further treatment.

7. Add the acid *very slowly* and *cautiously* to the boiling water, and continue to boil for another minute. Record any observation of note during this time. Remove the heat by turning off the gas flow to the burner. Use crucible tongs to remove the flask from the wire gauze. Set the flask aside for 10 min to allow the reaction, if any, to continue. Note and record any changes that occur.

8. Cool 20 mL of distilled water in an ice bath during this time.

9. When 10 min has elapsed, cool the flask in the ice bath. Filter the solid using suction. (This technique is described in the Introduction section of this manual.) Wash the solid with the ice-cold water. Continue drawing air through the solid for about 5 min to remove most of the water.

10. After 5 min, remove the filter paper and the solid from the funnel. Place the filter paper on a paper towel that is appropriately marked so that you will not forget the identity of this sample.

11. Repeat Steps 1, 2, and 5 through 10 using the other sample.

12. Wash your test tubes, rinse them with distilled water, and dry them.

13. Test the solubility of each filtered solid using a pea-sized portion of the solid and 2 mL of distilled water. Record your observations.

14. The identities of the samples should now be confirmed. Even though the samples have been partially dried, they are still too wet for their exact melting behavior to be meaningful.

    CAUTION: Before you leave the laboratory, make sure that your gas outlet and those of your neighbors are closed.

Date	______________	Student Name	______________
Course/Section	______________	Team Members	______________
Instructor	______________		______________

Geometric Isomers

Prelaboratory Assignment

1. Define the following terms:

   a. σ bond

   b. Single bond

   c. π bond

   d. Double bond

   e. Geometric isomers

   f. Isomerization

2. a. What geometric isomers will be examined during this experiment? Draw their structures and give their names.

b. Why are these isomers different compounds?

c. What prevents these isomers from being only one compound?

d. How do the properties of these isomers differ?

Student name: _______________ Course/Section: ________________ Date: ______________

3. What safety precautions are to be followed in this experiment?

Date ______________ Student Name ______________
Course/Section ______________ Team Members ______________
Instructor ______________ ______________

Geometric Isomers

Results

	Sample 1	Sample 2
Behavior on heating		
Behavior in water		
Behavior during attempted isomerization		
Behavior in water after attempted isomerization		

Questions

1. Identify each sample. What data supports your reasoning in making your identification?

2. Although maleic and fumaric acids are geometric isomers, *cis* and *trans* isomers of succinic acid (Figure 10.1(C)) do not exist. Explain why this is true.

3. Build molecular models of maleic, fumaric, and succinic acids (optional).

   a. Do the models show why these are not identical compounds? Why or why not?.

   b. Compare the rotation, if any, about the carbon–carbon double bonds in maleic and fumaric acids and about the corresponding carbon–carbon single bond in succinic acid.

Student name: ________________ Course/Section: _________________ Date: _______________

c. Why are *cis* and *trans* isomers of succinic acid impossible?

d. Use your models to examine isomerization. What must happen during isomerization?

11. A Student's View of Liquids and Solids

Introduction

A student of general chemistry is usually asked to carry out experiments for which the methods and procedures have been outlined in detail. Very little latitude is granted. This experiment, which focuses on certain properties of liquids and solids (Ebbing/Gammon, Chapter 11), provides the opportunity for students to develop their own experimental procedure to measure two physical properties.

Purpose

There are two principal tasks in this experiment. First, you will measure the densities of one or more solids whose crystalline lattices belong to the cubic system. Second, you will determine the heat (enthalpy) of fusion for ice.

Intermolecular Forces

Most, if not all, of the differences among the properties of the three states of matter are due to differences in their intermolecular forces. These forces are largest for solids, somewhat diminished for liquids, and smallest for gases. As you might expect, the densities of solids, liquids, and gases and the enthalpies that accompany phase changes reflect these forces and their differences.

Concept of the Experiment

This experiment gives you a unique chance to plan and execute procedures that will enable you to accomplish the two tasks described in the following paragraphs. No unusual glassware or equipment is required.

Your first task is to measure the densities of two or more metals. These metals may be powders, irregularly shaped pieces, or pieces with well-defined shapes. You will need to devise a general method for determining density that will not be affected by the form or shape of the metal.

FIGURE 11.1

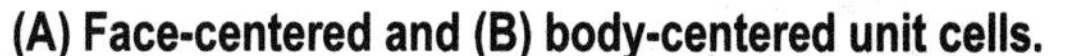

(A) Face-centered and (B) body-centered unit cells.

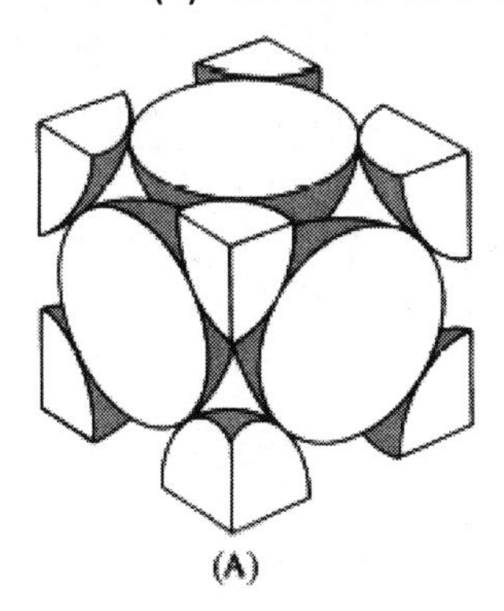

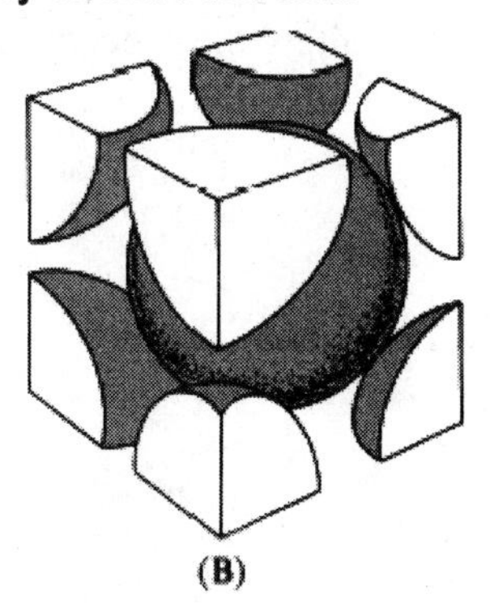

These metals will have either a face-centered cubic unit cell or a body-centered cubic unit cell. A unit cell is the smallest box-like arrangement of atoms and parts of atoms from which you can imagine creating a crystal by stacking them one upon the other or one next to the other. Both unit cells are shown in Figure 11.1. When coupled with the edge length of the unit cells, this information will enable you to calculate the densities of these metals. You will then be able to compare the densities you have measured with the ones you have calculated. You will also calculate the empty space in each lattice and show that it is independent of the length of the edge of the unit cell.

Your second task is to determine the heat (enthalpy) of fusion (Ebbing/Gammon, Section 11.2) for ice. The calorimeter will consist of two nested Styrofoam (polystyrene) coffee cups. This calorimeter will already be familiar to you if you did the experiment "Thermochemistry and Hess's Law." To become more familiar with it, read Appendix B carefully. It may give you some ideas about approaching this part of the experiment. Ice and, of course, distilled water will be available.

In addition to creating the procedures for this task, you will also be required to develop a mathematical equation that will allow you to calculate the desired enthalpy. Use common sense when you construct this equation. Remember that heat will flow from warm water and a warm calorimeter to ice and cold water.

Procedure

Getting Started

1. You may be asked to work with a group of students. If so, your laboratory instructor will designate the size of the group and, possibly, its members.
2. Determine and record the temperature of the laboratory.
3. Obtain samples of two or more metals and the required unit cell structural information for each sample. Record that information below.

Metal	**Structure***	**Edge Length (Å)**
____________	____________	____________
____________	____________	____________
____________	____________	____________

***FCC = face-centered cubic; BCC = body-centered cubic.**

Doing the Experiment

1. To do the experiment, follow the procedures that you outlined in the Prelaboratory Assignment.

Date ______________________ Student Name ______________________
Course/Section ______________________ Team Members ______________________
Instructor ______________________ ______________________

A Student's View of Liquids and Solids

Prelaboratory Assignment

1. Give the step-by-step procedures that you will use for the following tasks:

   a. Measuring the density of the metals

   b. Measuring the heat of fusion of ice

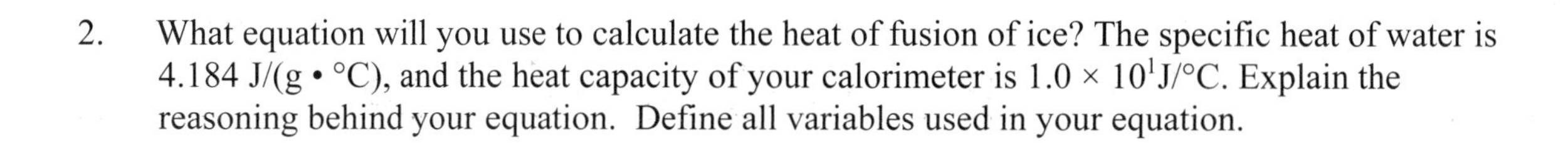

2. What equation will you use to calculate the heat of fusion of ice? The specific heat of water is 4.184 J/(g • °C), and the heat capacity of your calorimeter is 1.0×10^1 J/°C. Explain the reasoning behind your equation. Define all variables used in your equation.

3. What special safety precautions, if any, will be necessary in following your procedures?

Date	____________	Student Name	____________
Course/Section	____________	Team Members	____________
Instructor	____________		____________

A Student's View of Liquids and Solids

Results

Student name: ________________ Course/Section: ________________ Date: ________________

Questions

1. Calculate the percentage of empty space in 1 mol of water at 25°C. Obtain the density from the experiment "Some Measurements of Mass and Volume." The volume of a molecule of water can be taken as the sum of the volumes of the two hydrogen atoms and the oxygen atom. Use the single-bond covalent radii of 37 pm for hydrogen and 66 pm for oxygen and $V = 4\pi r^3/3$ to calculate the volume of each atom.

2. a. Calculate the density of each metal from its structure, edge length, and atomic weight.

b. In a few sentences, compare the calculated and experimental densities of each metal.

3. a. Calculate the percentage of empty space in a face-centered cubic lattice, and show that it does not depend on the edge length of the unit cell or on the size of the atoms in the unit cell. For this calculation, you must find the edge length in terms of the radii of the atoms in the unit cell, the total volume of the unit cell in terms of the edge length, and the filled volume from the total volume of the atoms in the unit cell.

Student name: ________________ Course/Section: ________________ Date: ______________

b. Calculate the percentage of empty space in a body-centered cubic lattice, and show that it does not depend on the edge length or the size of the atoms in the unit cell.

c. Provide a general comparison of the empty space in a solid with that in a liquid. If you did the experiment "Boyle's Law and the Empty Space in Air," include the empty space in a gas in your comparison.

4. Compare your experimental value for the heat of fusion of ice with the accepted value, 6.01 kJ/mol. Comment on the comparison, and try to explain your error (if it exists). How would you change your procedure to improve its accuracy?

12A. A Molar Mass from Freezing-Point Depression

Introduction

Solutions have many useful characteristics (Ebbing/Gammon, Chapter 12). Among these are the *colligative properties.* These properties, which include freezing-point depression, depend only on the number of solute molecules in a given quantity of solvent. They do not depend on the physical properties of the solute. As a consequence, colligative properties can be used to determine the molar mass of a solute (Ebbing/Gammon, Section 12.6).

Purpose

You will measure the freezing-point depressions that occur with solutions containing a solute whose empirical formula is known. Your data and those of the entire class will be used to calculate the molar mass of the solute and its molecular formula.

Concept of the Experiment

The apparatus you will use will probably resemble that shown in Figure 12A.1. The solvent will be cyclohexane. The solute will be a substance whose empirical formula is C_3H_2Cl.

You will determine the freezing-point depression, ΔT_f. This quantity is proportional to c_m, the *molal* concentration or *molality* of the solute:

$$\Delta T_f = K_f c_m$$

The *molality* of a solution defined as the number of moles of solute dissolved in one kilogram of solvent. The proportionality constant, K_f, is called the freezing-point-depression constant. It has a value of 20.5°C/*m* for cyclohexane. You can calculate the molar mass of a solute from ΔT_f, K_f, the mass of the solute, and the mass of the solvent (Ebbing/Gammon, Example 12.12).

In order to determine ΔT_f, you will measure the freezing point of pure cyclohexane and the freezing point of a solution of the solute in cyclohexane. You should use the following methods to obtain these temperatures.

FIGURE 12A.1

An apparatus suitable for determining the freezing point of a solvent or a solution.

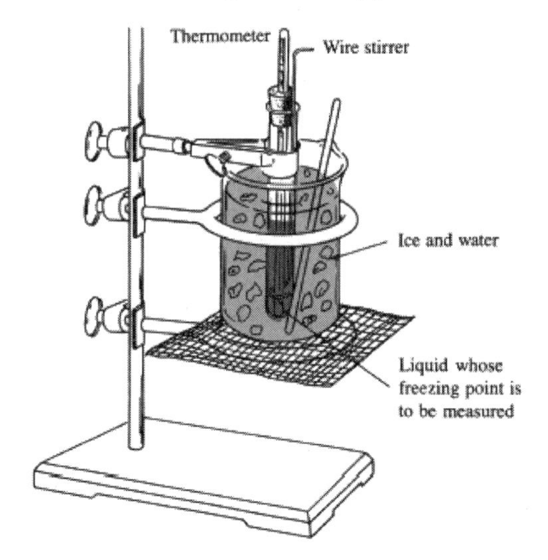

To evaluate the freezing point of pure cyclohexane, you will cool a sample of this substance and measure the temperature as a function of time. The cooling curve in Figure 12A.2 shows some typical results. As cooling begins, the temperature of the cyclohexane decreases rapidly. However, as soon as the solution starts to freeze, the temperature no longer decreases but remains constant until all the cyclohexane has frozen. When that has occurred, the temperature again decreases. The freezing point is the constant temperature that occurs while cyclohexane is freezing.

FIGURE 12A.2

A typical cooling curve for pure cyclohexane. The freezing point is t_f, the constant temperature in the middle of the cooling curve.

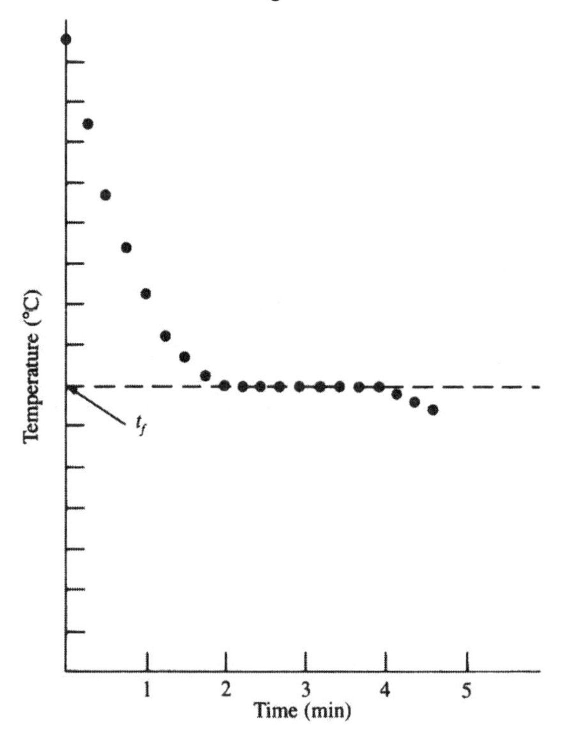

The freezing point of a solution of solute in cyclohexane can be obtained in a similar way, but there is a difference. This difference can be seen in Figure 12A.3. Cooling again results in an initial rapid decrease in the temperature of the solution. However, with the solution, unlike pure cyclohexane, the temperature does not remain constant until all of the solvent has frozen. As the cyclohexane in the solution freezes, the solution becomes more concentrated because less liquid remains. The molality of the solute increases and the freezing point must decrease still further. The result is a steadily decreasing freezing temperature rather than a constant freezing temperature. When all the solvent has frozen, the temperature decreases more rapidly.

The freezing point that you want is the initial freezing point, because that is the only point at which you will know how much liquid solvent remains. You can obtain the initial freezing point by drawing two solid lines that pass through most of the data, as shown in Figure 12A.3. The point at which these lines intersect corresponds to the initial freezing point.

FIGURE 12A.3

A typical cooling curve for a solution composed of a solute dissolved in cyclohexane. Two straight lines have been drawn to describe the data immediately before and during freezing. The intersection of these lines occurs at t_f', the initial freezing point of the solution. Using the results from Figure 12A.2, ΔT_f is given by $t_f - t_f'$.

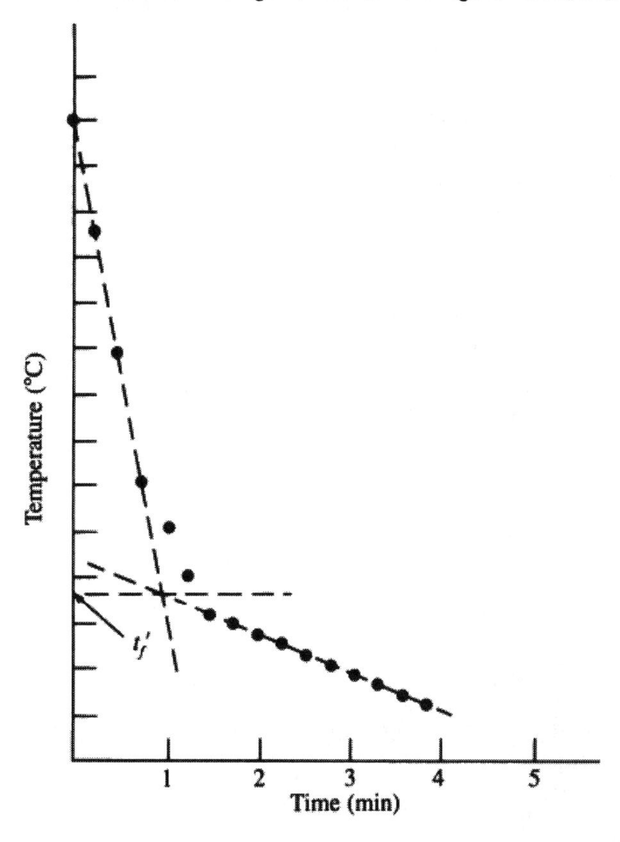

FIGURE 12A.4

A cooling curve obtained from an inefficiently stirred solution. Compare this curve with that shown in Figure 12A.3.

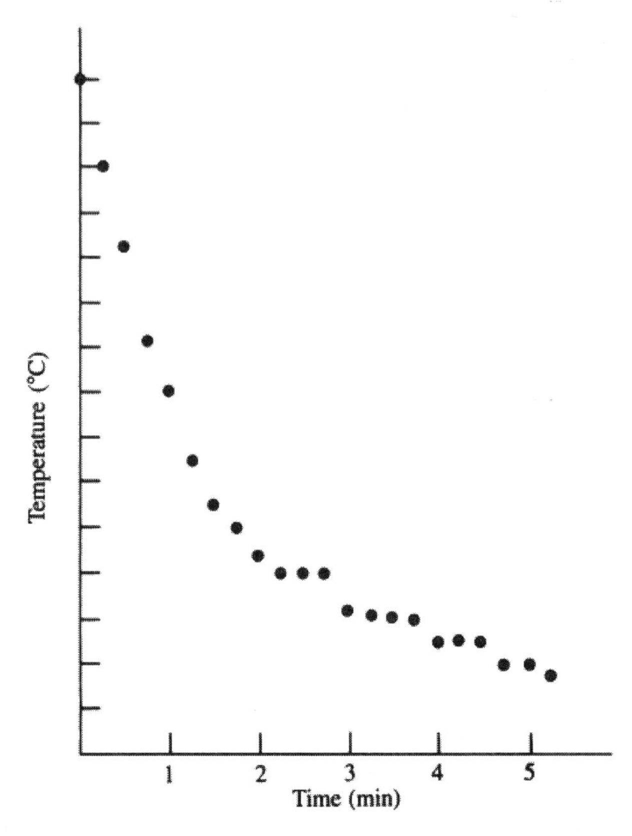

You will not obtain good cooling curves, such as those shown in Figures 12A.2 and 12A.3, unless you maintain efficient mixing of your solution by stirring constantly. If you do not stir the mixture well, you will obtain a series of stair steps instead of a smooth decrease in the temperature. An example is shown in Figure 12A.4. Drawing a line to describe this kind of data is very difficult, and the ultimate result is not very accurate.

Procedure

Getting Started

1. Your laboratory instructor will probably ask you to work with a partner.
2. Obtain directions for discarding the cyclohexane solution that you will use in this experiment from your laboratory instructor.
3. Obtain the apparatus required for this experiment. It will have been described or demonstrated by your laboratory instructor.
4. Make sure the apparatus is clean and dry.
5. Remember the following safety precaution whenever you use cyclohexane during this experiment:

CAUTION: Cyclohexane is flammable. No open flames are allowed during this experiment.

Measuring the Freezing Point of Cyclohexane

1. Pipet exactly 20.0 mL of cyclohexane into the apparatus. Close it immediately. Evaporation of this volatile substance must be avoided for best results.

   CAUTION: Never use your mouth to draw a liquid into the pipet. Use a rubber suction bulb or some other suction device.

2. Calculate and record the mass of cyclohexane. Use a density of 0.779 g/mL.
3. Place the apparatus in a beaker that contains crushed ice and water.
4. Stir the cyclohexane gently but constantly.
5. After the temperature has dropped to about 14–15°C, begin recording the temperature to the nearest 0.1°C every 15 s. A total time of 5–10 min will probably be required.
6. After the cyclohexane has frozen and the temperature has begun to decrease again, remove the apparatus from the beaker. Allow the cyclohexane to melt completely.
7. Repeat Steps 3 through 6.
8. Plot the data, find the freezing point in each case using the method described in Figure 12.A.2, and calculate the mean freezing point.
9. *Save* the cyclohexane for the next part of the experiment.

Measuring the Freezing Points of the Solutions

1. Obtain 2 pieces of waxed weighing paper. Mark each one for recognition.
2. Obtain and record the mass of one of these, using your most precise balance. Place 0.24–0.25 g of the solute directly on the paper. Obtain and record the mass of the paper and solute. Calculate the mass of the solute.
3. Repeat for the other piece of weighing paper. However, use only 0.10–0.11 g of the solute on this paper.
4. Transfer the first sample into the apparatus.
5. Stir until you obtain a clear solution. All of the solid, including any on the walls of the apparatus, must dissolve before the measurement can begin.
6. Stir the solution gently but constantly.
7. Cool the solution to about 14–15°C, and then record the temperature at 15-s intervals. Continue cooling until the entire solution is frozen.
8. Remove the apparatus from the beaker and allow the solution to melt completely.
9. While you are waiting, plot your data and look at the resulting graph. Has the temperature fallen smoothly? If you have a stair- step effect, repeat Steps 6 through 8 with faster stirring.
10. Add the second sample to the solution. The mass of the solute in solution will now be the combined masses of the first and second samples.
11. Repeat Steps 6 through 9.
12. Plot the data for each sample to find the freezing points.
13. Calculate each ΔT_f using the method outlined in Figure 12.A.3
14. Calculate the molar mass (M_m) of the solute from each ΔT_f. Report your answer to an appropriate number of significant figures.

15. Obtain the mean molar mass.
16. Share your results with the rest of the class and obtain their results. Calculate the grand average.
17. Calculate the molecular formula of the solute from the grand average and the empirical formula (C_3H_2Cl).

Date ______ Student Name ______
Course/Section ______ Team Members ______
Instructor ______ ______

A Molar Mass from Freezing-Point Depression

Prelaboratory Assignment

1. Provide definitions for the following terms:

   a. Solution

   b. Solute

   c. Solvent

   d. Colligative property

   e. Freezing-point depression

2. a. What is the objective of this experiment?

   b. How will that objective be achieved?

3. A 0.2436-g sample of an unknown substance was dissolved in 20.0 mL of cyclohexane. The density of cyclohexane is 0.779 g/mL. The freezing-point depression was 2.5°C. Calculate the molar mass of the unknown substance.

4. What safety rules must be observed during this experiment?

Date ______________________ Student Name ______________________

Course/Section ______________________ Team Members ______________________

Instructor ______________________ ______________________

A Molar Mass from Freezing-Point Depression

Results

1. Volume of cyclohexane: __________mL

   Mass of cyclohexane: __________g, or__________kg

   Calculations:

2. *Freezing point of cyclohexane*

Trial 1		Trial 2	
Time	**Temperature (°C)**	**Time**	**Temperature (°C)**

Freezing point of cyclohexane: ____________ First Trial

____________ Second Trial

____________ Mean

Student name: ______________ Course/Section: ______________ Date: ______________

3. *Freezing points of the solutions*

Sample	1	2
Mass of paper + solute (g)	______	______
Mass of paper (g)	______	______
Mass of solute (g)	______	______

Sample 1		Sample 1 + 2	
Time	Temperature (°C)	Time	Temperature (°C)
______	______	______	______
______	______	______	______
______	______	______	______
______	______	______	______
______	______	______	______
______	______	______	______
______	______	______	______
______	______	______	______
______	______	______	______
______	______	______	______
______	______	______	______
______	______	______	______
______	______	______	______
______	______	______	______
______	______	______	______
______	______	______	______
______	______	______	______
______	______	______	______
______	______	______	______
______	______	______	______

Freezing points of solutions:

Sample 1:

____________°C; ΔT_f = ____________°C; M_m = ____________

Samples 1 + 2:

____________°C; ΔT_f = ____________°C; M_m = ____________

Mean M_m: ____________

Calculations:

Student name: ________________ Course/Section: ________________ Date: ________________

Record M_m from your classmates (include your value).

______	______	______	______	______	______
______	______	______	______	______	______
______	______	______	______	______	______
______	______	______	______	______	______

Grand average: ____________

Molecular formula:

Calculations:

Question

1. What would be the effect of each of the following on the calculated molecular weight of the solute? Think carefully!

   a. Some cyclohexane evaporated while the freezing point of *pure* cyclohexane was being measured.

b. Some cyclohexane evaporated after the solute was added.

c. A foreign solute was already present in the cyclohexane.

d. The thermometer is not calibrated correctly. It gives a temperature that is 1.5°C too low at all temperatures.

12B. Softening Hard Water

Introduction

Our water comes from lakes, rivers, and wells. Even if it is fit to drink, it is never chemically pure; it is a solution. Water from these sources always contains a variety of dissolved ions, including Ca^{2+}, Mg^{2+}, Fe^{3+}, HCO_3^-, and SO_4^{2-} ions. Water with relatively small concentrations of Ca^{2+} and Mg^{2+} ions is called *soft water*. If the concentrations of these ions are relatively large, the water is called *hard water*.

Hard water causes many problems including the formation of "scale or hard water spots" which are insoluble salts deposited on the inside of hot water heaters, tubs and showers. However, there are several methods for removing Ca^{2+} and Mg^{2+} ions. Any method that removes these ions is called *water softening*. One technique for softening water is *ion exchange.*

Purpose

This experiment begins with an examination of some reactions of aqueous Ca^{2+} and Mg^{2+} ions. Some of these reactions are important in our natural waters. The experiment concludes with a study of water softening by ion exchange.

The Source of Hard Water

Calcium and magnesium ions are often found in minerals as the carbonates $CaCO_3$ and $MgCO_3$. These substances are insoluble in pure water. Hard water is a result of their solubility in the dilute solutions of carbonic acid (H_2CO_3) that occur in natural water. Atmospheric carbon dioxide (CO_2) dissolves in natural water to form this acid according to the equation

$$CO_2(g) + H_2O(l) \rightarrow H_2CO_3(aq)$$

The carbonic acid formed in this process reacts with a carbonate mineral to give a soluble substance. For example, the acid reacts with $CaCO_3$ to form soluble calcium hydrogen carbonate which dissociates in water to form calcium and bicarbonate ions:

$$CaCO_3(s) + H_2CO_3(aq) \rightarrow Ca^{2+}(aq) + 2HCO_3^-(aq)$$

The overall equation for this process is the sum of these equations:

$$CaCO_3(s) + CO_2(g) + H_2O(l) \rightarrow Ca^{2+}(aq) + 2HCO_3^-(aq)$$

A similar reaction equation can be written for $MgCO_3$.

The Effects of Hard Water

The reactions that are responsible for the solubilities of $CaCO_3$ and $MgCO_3$ in dilute solutions of H_2CO_3 are readily reversed when heat is applied. The reversal occurs because CO_2 is evolved from the hot solutions. As a result, the solution becomes less acidic, and the insoluble carbonates return. These carbonates are partly responsible for the residues that coat the inside surfaces of tea kettles, hot-water pipes, boilers, and heat exchangers.

Another problem occurs when soap is used in hard water. The Na^+ ion in sodium stearate ($C_{17}H_{35}COO^- Na^+$), a common ingredient in soap, is replaced by Ca^{2+} and Mg^{2+} ions to form an insoluble cottage cheese curd-like scum. This material, which consists of calcium and magnesium stearates, is responsible for the dull appearance of clothes washed in hard water and for the ring of scum that can form on a bathtub. Detergents do not form insoluble calcium and magnesium salts and have replaced soap to avoid these effects.

Measuring Hardness

Hardness is usually measured through a titration of the water with a solution of the sodium salt of dihydrogen ethylenediaminetetraacetate ions. Figure 12B.1 shows the structure of the parent anion, which is called EDTA. The doubly protonated EDTA anion that is used in the titration reacts with a Ca^{2+} ion (or almost any other metal ion) to form a complex ion (Ebbing/Gammon, Section 23.3). The equation for this reaction is

$$H_2EDATA^{2-}(aq) + Ca^{2+}(aq) \rightleftharpoons Ca(EDTA)^{2-}(aq) + 2H^{+}(aq)$$

A similar reaction occurs with Mg^{2+} ions.

The endpoint in the titration can be detected with an indicator such as Eriochrome Black T. This indicator, whose aqueous solutions are blue, forms rose-pink complex ions with Ca2+ and Mg2+ ions. For example, the reaction with Ca2+ ions can be written as

$$\underset{\text{}}{Ca^{2+}(aq)} + \underset{\text{(blue)}}{In(aq)} \rightleftharpoons \underset{\text{(rose pink)}}{Ca(In)^{2+}(aq)}$$

where In represents the complicated chemical formula of the indicator. When H_2EDTA^{2-} anions are added to this solution during a titration, the metal ions are removed from the indicator complex ion to form the complex ion with EDTA. The reaction with Ca^{2+} ions is

$$H_2EDTA^{2-}(aq) + \underset{\text{(rose pink)}}{Ca(In)^{2+}(aq)} \rightleftharpoons Ca(EDTA)^{2-}(aq) + \underset{\text{(blue)}}{In(aq)} + 2H^{+}(aq)$$

The reaction occurs because the $Ca(EDTA)^{2+}$ ion is more stable than the indicator complex ion. A similar reaction can be written for magnesium. The endpoint of the titration occurs when all of the indicator has released the calcium ions and the blue color appears.

FIGURE 12B.1

The structure of the ethylenediaminetetraacetate (EDTA) anion.

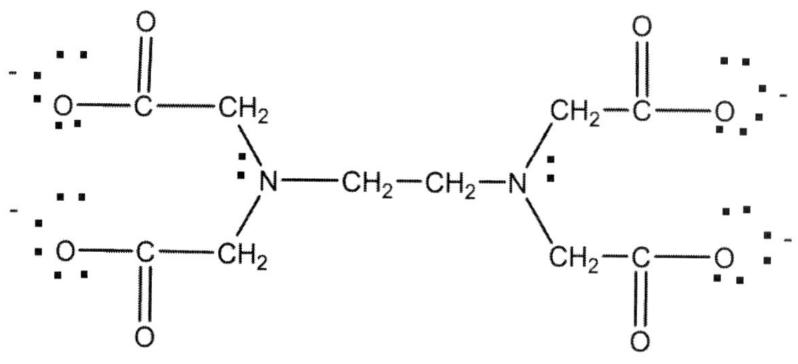

Ion Exchange

A cation-exchange resin will be used in this experiment. This resin is an insoluble organic substance whose molecules consist of long chains of atoms. The sulfonic acid group, –SO3H, is chemically bonded to the chain in many places. Figure 12B.2 shows the mechanism for cation exchange. In ion exchange, one type of ion simply replaces an ion of another type.

FIGURE 12B.2

A cation-exchange resin in the H^+ form (center), the Ca^{2+} form (left), and the Na^+ form (right).

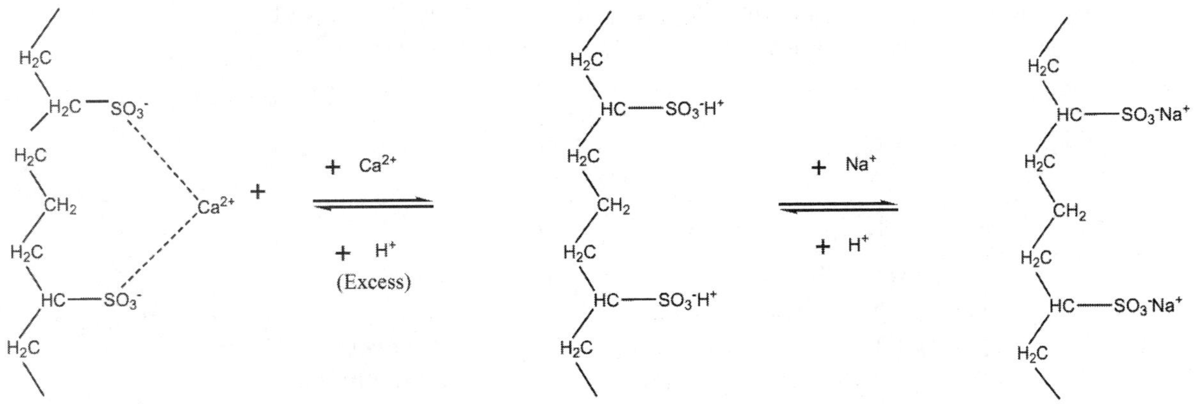

Ions with a +2 charge, such as Ca^{2+} ions, are more strongly bound to the resin than are ions with a +1 charge, such as Na^+ or H^+ ions. As a result, a cation-exchange resin in the H^+ form will easily exchange ions with a dilute solution of Ca^{2+} and Mg^{2+} ions. After this exchange has occurred, the resin in its original H^+ form can be regenerated, but a large excess of a strong acid must be used. Ion-exchange resins are best employed in columns so that a solution can percolate or move slowly through the resin.

Concept of the Experiment

Because $CaCO_3$ and $MgCO_3$, along with CO_2, play important roles in the formation of hard water and in some of its effects, the first part of the experiment will focus on these substances.

Water softening by ion exchange is the subject of the second part of this experiment. Although ion-exchange resins in columns are very efficient, this experiment uses an easier technique. This method consists of stirring the resin and a water sample in a beaker. You should still achieve quantitative or near-quantitative cation exchange, because you will use a large excess of the resin.

To judge the hardness of the water sample before and after ion exchange, you will use Na_2H_2EDTA. However, you will use a semi-quantitative titration via a medicine dropper. The sample will be tap water or, if your water is very soft, an artificial sample provided by your laboratory instructor.

Procedure

Getting Started

1. Obtain 4 small test tubes and 2 pieces of blue litmus paper. Additional litmus paper can be obtained if you need it.

2. Obtain about 6 g of dry cation-exchange resin or about 9 g of wet resin and place it into a 250-mL beaker.

   This resin is very expensive, so you will be requiredto recycle it when you have finished the experiment. Your laboratory instructor will provide instructions on where to recycle your resin sample.

3. Your laboratory instructor will tell you whether you are to use tap water or an artificial sample for the ion exchange part of this experiment.

Observing Reactions of Ca^{2+} and Mg^{2+} Ions

1. Add 10 drops of 0.1 *M* $Ca(NO_3)_2$ to each of 3 test tubes.

2. Add 10 drops of 0.1 *M* Na_2CO_3 to the first test tube, another 10 drops of this solution to the second test tube, and 10 drops of 0.1 *M* $NaHCO_3$ to the third test tube.

3. Shake each of the test tubes gently. Observe the results carefully and record them.

4. Add dry ice (solid CO_2) slowly to the first test tube. You should use a quantity about the size of 2 peas. Shake the test tube gently until no further effects can be seen. Record the result.

5. Add 2 drops of 6 *M* HCl to the second test tube. Record the result.

   CAUTION: Handle the solution of hydrochloric acid carefully. It can cause chemical burns, in addition to ruining your clothing. If you spill any on you, wash the contaminated area thoroughly and report the incident to your laboratory instructor. You may require further treatment. Dry ice is extremely cold. Wear protective cloth or leather gloves when handling it. Do not handle the dry ice with your bare hands as prolonged exposure could result in damage to skin cells.

6. Heat the third test tube gently in the flame of your laboratory burner.

   CAUTION: Do not point the test tube toward anyone. Do not let the flame linger in any one place. Move the test tube continuously in the flame.

   Record the result.

7. Cool the third test tube in tap water. Add 2 drops of 6 *M* HCl. Record the result.

8. Add a pea-sized portion of solid $CaCO_3$ to the fourth test tube. Add several drops of 6 *M* HCl and observe. Record the result. What you have seen here also occurred in Steps 5 and 7, but it may not have been noticed.

9. Discard the solutions as directed by your laboratory instructor. Wash and rinse the test tubes.

10. Repeat Steps 1 through 7 using 0.1 *M* $Mg(NO_3)_2$ instead of 0.1 *M* $Ca(NO_3)_2$.

Preparing the Resin

1. You need to make sure that the resin is in the H+ form, even if you think you obtained it in this form.

2. Add 6 *M* HCl to the beaker containing the resin until the resin is barely covered. Allow the mixture to stand for about 2 min.

3. Add about 200 mL of distilled water, and let the mixture stand until the resin settles to the bottom of the beaker.

4. *Carefully* decant (pour off) the solution above the resin until only a few milliliters of solution remain.

5. Wash the resin by adding 200 mL of distilled water. Let the resin settle to the bottom, and then decant again.

6. Repeat Step 5 three more times. Discard all of the wash water as directed by your laboratory instructor.

7. Repeat Step 5 again, but test the solution with blue litmus paper before decanting. If the paper turns pink, continue washing according to Step 5 until no change occurs when you test the

solution with blue litmus paper.

Softening Hard Water

1. Cover the resin in the beaker with a quantity of tap water or the artificial sample. Use 100 mL unless your laboratory instructor provides other directions. If you are using an artificial sample, you will need enough for water softening (100 mL) plus an additional quantity of about 50 mL.
2. Let the mixture of resin and water stand for 10 min with occasional stirring.
3. During this time, add 20 mL of unsoftened water (either tap water or your artificial sample) to a clean 125-mL Erlenmeyer flask from a clean graduated cylinder.
4. Add 5 mL of an NH_3–NH_4Cl buffer ($[OH^-] \approx 10^{-4}$ *M*) to this flask.
5. Add a very small portion of a solid mixture of Eriochrome Black T and sodium chloride. This portion should not exceed ⅛ inch on the tip of a metal spatula. Hard water should provide a light rose-pink color.
6. Add 0.01 *M* Na_2H_2EDTA *slowly* by drops from a medicine dropper, counting the drops and swirling the solution in the flask after each drop is added. As the endpoint is approached, you should see a lavender color. After this point, add the solution more slowly until the first appearance of the blue color. Record the number of drops that were required.
7. If you wish, repeat Steps 3 through 6 with another 20 mL of unsoftened water. Calculate and record the mean.
8. After 10 min has elapsed, *carefully* decant the softened water (from step 2) into a clean beaker leaving only a few milliliters of the water remaining with the resin.
9. Repeat Steps 3 through 6 using softened water instead of unsoftened water.
10. The blue color will appear immediately if virtually all of the Ca^{2+} and Mg^{2+} ions in the water have been removed. If so, record this result. If not, repeat Step 6.

Finishing the Experiment

1. Convert the resin to the H^+ form, using the same method that you used to prepare the resin.
2. Return the resin to your laboratory instructor.

CAUTION: Before you leave the laboratory, make sure that your gas outlet and those of your neighbors are closed.

Date ______________ Student Name ______________
Course/Section ______________ Team Members ______________
Instructor ______________ ______________

Softening Hard Water

Prelaboratory Assignment

1. Provide definitions for the following terms:

   a. Hard water

   b. Soft water

   c. Water softening

   d. Ion exchange

   e. Cation-exchange resin

2. a. Describe two detrimental effects that result from using hard water.

b. Write balanced chemical equations that describe the insoluble curd-like precipitate that is formed when soap is used in hard water.

3. Write balanced chemical equations that describe the titration of Mg^{2+} ions with $H_2EDTA2-$ ions in the presence of the indicator.

4. What will you do with the cation-exchange resin after the experiment?

5. What safety precautions should be observed during this experiment?

Date		Student Name	
Course/Section		Team Members	
Instructor			

Softening Hard Water

Results

1. *Reactions of Ca^{2+} and Mg^{2+} ions*

	$Ca(NO_3)_2$	$Mg(NO_3)_2$
Na_2CO_3		
$Na_2CO_3 + CO_2$		
$Na_2CO_3 + HCl$		
$NaHCO_3$		
$NaHCO_3$ + heat		
$NaHCO_3$ + heat + HCl		

Solid $CaCO_3 + HCl$	

2. *Softening hard water*

Drops of 0.01 *M* Na_2H_2EDTA

Before ion exchange: ________________

After ion exchange: ________________

Questions

1. Write a balanced chemical equation for every reaction that you observed with Ca^{2+} and Mg^{2+} ions in the first part of the experiment. If Mg^{2+} ions behaved somewhat differently from Ca^{2+} ions, give a reason.

2. Determine the percent efficiency of your water softening from the number of drops of the Na_2H_2EDTA solution that were required before and after you treated the water with the cation-exchange resin.

Student name: _______________ Course/Section: ________________ Date: ______________

3. Describe how a cation-exchange resin and a solution of NaOH with a known concentration could be used to determine the amount of NaCl in an aqueous solution.

13. The Rate of an Iodine Clock Reaction

Introduction

Some reactions, including most of the ones that you have seen in this manual, occur so rapidly that they are over as soon as the reactants are mixed. There are many more reactions, however, that are slower; they may require minutes, hours, days, or even years to reach completion. A few chemical reactions are so slow that it is difficult to show that they happen at all.

Why are there wide differences in reaction rates? They are largely due to the differing molecular characteristics of the reactants. However, there are other factors that also affect the rate of any given reaction (Ebbing/Gammon, Chapter 13). Three of these factors are the concentrations of the reactants, the concentration of a catalyst, and the temperature at which the reaction occurs.

Purpose

In the quantitative part of this experiment, you will determine the rate law for an iodine clock reaction and the influence of temperature on that reaction. In the qualitative part, you will evaluate the effect of a catalyst on the rate of this reaction.

What is an Iodine Clock Reaction?

There are actually several different iodine clock reactions. All of them, however, have a common feature: The completion of any one of them is signaled by the *sudden* appearance of the dark color that is characteristic of the interaction of molecular iodine (I_2) with starch. When the reaction is handled correctly, this color appears so abruptly that it can be as startling as the sudden sound of the alarm bell of a clock.

The rate of an iodine clock reaction depends on the concentrations of the reactants, as you would expect. As a result, the time required for the appearance of the dark color can be adjusted by adjusting the concentrations of the reactants. Hence it is possible to set the color alarm on a clock reaction just as you can set the sound alarm on a clock.

Concept of the Experiment

You will study the rate of the reduction of potassium persulfate ($K_2S_2O_8$) with sodium iodide (NaI). The net ionic equation for this reaction is

$$S_2O_8^{2-} + 2I^- \rightarrow 2SO_4^{2-} + I_2$$

One of the purposes of this experiment is to determine the rate law for this reaction. The rate law's general form will be (Ebbing/Gammon, Section 13.3)

$$\text{Rate} = k[S_2O_8^{2-}]^q[I^-]^r$$

and you will need to determine values for the rate constant k and the exponents q and r. Remember that the exponents must be determined experimentally; they cannot be obtained simply by looking at the balanced equation.

You will use three experiments and the initial-rate method (Ebbing/Gammon, Section 13.3) to determine the rate law. Experiment 1 will be the reference with which the other two will be compared. In Experiment 2, the initial concentration of the $S_2O_8^{2-}$ ions will be twice that in Experiment 1, but the concentration of I^- ions will be unchanged. The general form of the rate law for each of these experiments will be

Experiment 1: Rate$_1$ = $k[S_2O_8^{2-}]^q[I^-]^r$

Experiment 2: Rate$_2$ = $k(2[S_2O_8])$ [I] = 2 × rate$_1$

The rate in the second experiment will be 2^q times the original rate. In Experiment 3, the initial concentration of $S_2O_8^{2-}$ ions will be the same as that in Experiment 1, but the concentration of I^- ions will be doubled. The rate law will be

Experiment 3: Rate$_3$ = $k[S_2O_8^{2-}]^q(2[I^-])^r = 2^r \times$ rate$_1$

The rate in this experiment will be 2^r times the original rate.

You can determine the true values of q and r from the ratios rate$_2$/rate$_1$ and rate$_3$/rate$_1$. After you have evaluated the exponents, you will know the general form of the rate law, but the value of the rate constant will be unknown. You can obtain it from the rate law and any one of the three rates. Each rate should yield the same rate constant within experimental error.

The rate of the reaction is given by

$$\text{Rate} = -\frac{\Delta[S_2O_8^{2-}]}{\Delta t}$$

therefore you will need to know or measure $\Delta[S_2O_8^{2-}]$, the initial change in the concentration of $S_2O_8^{2-}$ ions, and Δt, the time elapsed during that change, as well as the initial concentrations of $\Delta[S_2O_8^{2-}]$ and I^- ions. An easy way to obtain $\Delta[S_2O_8^{2-}]$ is by coupling another reaction to the one we are studying. The new reaction is the reduction of I_2 by $Na_2S_2O_3$ (sodium thiosulfate). The net ionic equation for this reaction is

$$I_2 + 2S_2O_3^{2-} \rightarrow 2I^- + S_4O_6^{2-}$$

where $S_4O_6^{2-}$ is the tetrathionate ion. *It is important to remember that this reaction is used only as a means of studying the rate of the first reaction.*

The new reaction is fast. As a result, I_2 is consumed in this reaction as fast as it is formed in the first reaction. Essentially, there can be no I_2 present as long as there are $S_2O_3^{2-}$ ions present. However, these ions are being consumed because of the second reaction. As soon as all the $S_2O_3^{2-}$$_3$ions have reacted, I_2 from the first reaction begins to accumulate in the solution. You can detect the sudden presence of I_2 by the sudden appearance of the dark color from the interaction of I_2 with starch, which will be used as an indicator.

You will know the change in the amount of the persulfate ion, $\Delta[S_2O_8^{2-}]$, at the instant that this dark color appears. How? As you will show in the Prelaboratory Assignment, the stoichiometry and relative rates of the coupled reactions require the following relationship between $\Delta[S_2O_8^{2-}]$ and $\Delta[S_2O_3^{2-}]$:

$$\Delta[S_2O_8^{2-}] = (\tfrac{1}{2})\Delta[S_2O_3^{2-}]$$

where

$$\Delta[S_2O_3^{2-}] = \text{initial } [S_2O_3^{2-}] - \text{final } [S_2O_3^{2-}]$$

Essentially all the $S_2O_3^{2-}$ ions will have reacted when the dark color appears, so the final concentration of these ions is essentially zero, and $\Delta[S_2O_3^{2-}]$ becomes

$$\Delta[S_2O_3^{2-}] = \text{initial } [S_2O_3^{2-}]$$

If we keep the initial concentration of $S_2O_3^{2-}$ ions at a small value the change in thiosulfate, $\Delta[S_2O_3^{2-}]$, will be small and $\Delta[S_2O_8^{2-}]$ will be even smaller. As a consequence, there will be little change in the concentration of the reactants during the elapsed time Δt. This is a necessary condition for the initial-rates method.

The time required for the appearance of the dark color is affected by the concentration of the reactants, as you would expect from the general form of the rate law. The time is also affected to a degree by the overall concentration of ions and their charges. This influence, which is well known, cannot be predicted from the rate law. You will maintain a constant concentration of ions and charges by adding, where required, ionic substances that will not react. Thus Experiment 1 has one-half the amount of $K_2S_2O_8$ found in Experiment 2. The deficiency is overcome by the addition of inert potassium sulfate, K_2SO_4. Similarly, Experiment 1 has one-half the amount of NaI found in Experiment 3. The addition of inert NaCl keeps the total concentration of ions and charges at a constant value.

To demonstrate the influence of the temperature on the rate of this reaction, you will repeat Experiment 1 at an increased temperature. The mathematics required here is straightforward. The rate law will have the same general form at the higher and lower temperatures. As a result, q and r will not change when the temperature is varied. A new rate constant, however, will need to be evaluated at the higher temperature. The activation energy for the reaction can be calculated via the Arrhenius equation (Ebbing/Gammon, Section 13.6).

Finally, you will repeat Experiment 1 one more time to explore the effect of a catalyst in qualitative terms. The catalyst will be the Cu(II) ions in $CuSO_4$. No mathematical treatment will be required.

Procedure

Getting Started

1. Work with a partner.
2. Obtain 3 large test tubes with 3 rubber stoppers, a Mohr pipet, and a thermometer.
3. Obtain directions for discarding the solutions that you will use in this experiment from your laboratory instructor.

Completing the First Series of Experiments

1. Note and record the temperature of the laboratory to the nearest tenth of a degree.
2. Mark each of the test tubes with an identification number (1, 2, and 3).
3. Rinse the pipet with two 1-mL portions of the NaI solution. Each of these portions should be discarded.
4. Pipet the required volume of the NaI solution into each test tube. These volumes can be found in Table 13.1.

Table 13.1

Volumes (mL) of Solutions

Experiment	0.20 *M* NaI	0.20 *M* NaCl	0.010 *M* $Na_2S_2O_3$	2% Starch	0.20 *M* K_2SO_4	0.20 *M* $K_2S_2O_8$
1	2.0	2.0	2.0	1.0	2.0	2.0
2	2.0	2.0	2.0	1.0	0	4.0
3	4.0	0	2.0	1.0	2.0	2.0

5. Rinse the pipet with distilled water.

6. Using the required amount of NaCl solution instead of the NaI solution, repeat Steps 3 through 5, as described in Table 13.1.

7. Using the required amount of $Na_2S_2O_3$ solution instead of the NaI solution, repeat Steps 3 through 5, as described in Table 13.1.

8. Using the required amount of starch solution instead of the NaI solution, repeat Steps 3 through 5, as described in Table 13.1.

9. Using the required amount of K_2SO_4 solution instead of the NaI solution, repeat Steps 3 through 5, as described in Table 13.1.

10. Read Steps 11 and 12 completely before doing them. If you do not perform the steps correctly, the experiment will not work properly.

11. Note and record the time just as you begin to pipet the required amount of the $K_2S_2O_8$ solution for Experiment 1 into test tube 1. Quickly insert a rubber stopper in the test tube. Mix the solution by inverting the test tube about 15 times while shaking it simultaneously. *A completely homogeneous solution is required.*

12. Note and record the exact time at which the dark color appears. Be alert. This color should appear suddenly and uniformly throughout the solution at one time. If it does not, you have not mixed the solution thoroughly enough. Calculate the elapsed time. Experiment 1 should require less than 4 min.

13. Repeat Steps 11 and 12 with test tubes 2 and 3, using the quantities of the $K_2S_2O_8$ solution specified for Experiments 2 and 3 in Table 13.1.

14. Wash, rinse, and dry the test tubes and rubber stoppers.

15. Perform a second trial by repeating Steps 3 through 14.

16. If the elapsed times of the two trials of an experiment differ by more than 10 s, repeat the experiment until the times are within this range.

Completing the Second Series of Experiments

1. Renumber the test tubes (4, 5, and 0).

2. Prepare a beaker of water whose temperature is about 35°C by adding hot water to cold water.

3. The volumes of the solutions used in Experiment 4 will be the same as those used in Experiment 1.

4. Pipet the required volumes of the NaI, NaCl, $Na_2S_2O_3$, starch, and K_2SO_4 solutions into test tube 4. Place this test tube in the water bath.

5. Place about 5 mL of the $K_2S_2O_8$ solution in the clean, dry test tube whose number is 0. Put this test tube in the water bath.

6. Allow each test tube to remain in the water bath for about 5 min. Add small amounts of hot water to the bath during this time to maintain the temperature at about 35°C.

7. Initiate the reaction by pipeting the required volume of the warm $K_2S_2O_8$ solution into the other test tube in the water bath. Note and record the time.

8. Insert a rubber stopper into the test tube, remove the test tube from the bath, and shake the test tube as vigorously as you can for a few seconds. Replace the test tube in the bath. Watch carefully for the color change..

9. Note the time at which the dark color appears. Remove the stopper, and measure and record the temperature of the solution to the nearest tenth of a degree. Calculate the elapsed time.

10. The volumes of the solutions used in Experiment (test tube) 5 will be the same as those used in Experiment (test tube) 1, but do not heat the solutions. Add 1 drop of 0.2 *M* $CuSO_4$ and shake the test tube gently before you initiate the reaction with the required volume of the $K_2S_2O_8$ solution.

Date ______________________ Student Name ______________________
Course/Section ______________________ Team Members ______________________
Instructor ______________________ ______________________

The Rate of an Iodine Clock Reaction

Prelaboratory Assignment

1. What factors affect the rate of any given reaction?

2. a. Write the balancedchemical equation for the reaction whose rate is being studied.

   b. Write the balanced chemical equation for the reaction that will enable you to know $\Delta[S_2O_8^{2-}]$.

   c. Why are these coupled reactions called an iodine clock reaction?

   d. What chemical interaction is responsible for the dark color?

e. Why does the dark color appear suddenly, rather than as a gradual darkening of the solution?

f. When the dark color appears, $\Delta[S_2O_8^{2-}] = (1/2)\Delta[S_2O_3^{2-}]$, where $\Delta[S_2O_3]$ equals the initial concentration of $Na_2S_2O_3$. Why is this true?

3. Consider the composition of the solutions in Table 13.1. In which experiment will the longest time probably elapse before the appearance of the dark color? Why?

Student name: ________________ Course/Section: __________________ Date: _______________

4. What catalyst will be used? In which experiment will it be used?

5. Give names for the following substances:

a. $K_2S_2O_8$

b. S_2O^{2-}

c. $Na_2S_2O_3$

d. $S_2O_3^{2-}$

Date ______________________ Student Name ______________________
Course/Section ______________________ Team Members ______________________
Instructor ______________________ ______________________

The Rate of an Iodine Clock Reaction

Results

Laboratory temperature: ____________°C

Experiment	Trial	Start	End	Elapsed Time (s)	
1	1	________	________	________	
	2	________	________	________	
2	1	________	________	________	
	2	________	________	________	
3	1	________	________	________	
	2	________	________	________	
4	1	________	________	________	Temp.: ________°C
5	1	________	________	________	

Questions

1. a. Calculate the initial reaction rates for Experiments 1, 2, and 3 from $\Delta[S_2O_8^{2-}]/\Delta t$. Use mean elapsed times. Use correct units.

   b. Obtain the reaction orders with respect to the $S_2O_8^{2-}$ and I^- ions. Remember that your results have been influenced by experimental error.

tudent name: _______________ Course/Section: _________________ Date: _______________

c. What is the correct rate equation?

d. Calculate the initial concentrations of the $S_2O_8^{2-}$ and I^- ions in Experiments 1, 2, and 3. Remember that dilution occurred when the solutions were prepared.

e. Calculate the rate constant for Experiments 1, 2, and 3, and obtain the mean rate constant for the three experiments. Express you answer using the correct units.

2. a. Calculate the rate constant at the higher temperature used in Experiment 4.

b. Calculate the activation energy for the reaction.

3. What do your results demonstrate about the effect of a catalyst on the rate of a reaction?

14A. Le Châtelier's Principle

Introduction

All chemical reactions proceed until equilibrium is reached, provided none of the reactants or the products are removed from the reaction mixture. Le Châtelier's principle describes what happens to an equilibrium after it has been disturbed (Ebbing/Gammon, Sections 14.7, 14.8, and 14.9).

Purpose

You will study the application of Le Châtelier's principle by seeing the effect of the addition of Fe^{3+} and SCN^- to an equilibrium mixture of Fe^{3+}, SCN^-, and $Fe(SCN)^{2+}$; the effect of the addition of an acid to an equilibrium mixture of Ni^{2+}, NH_3, and $Ni(NH_3)_6^{2+}$; the effect of the addition of an acid and a base on the equilibrium involving an indicator; the effect of acids and bases on the solubility of $Ca(OH)_2$ in water; and the effect of temperature on an equilibrium mixture of Co^{2+}, Cl^-, and $CoCl_4^{2-}$.

New Substances

In this experiment you will encounter some substances that you may not have seen before. The reaction between Fe^{3+} and SCN^- (thiocyanate ion) gives $Fe(SCN)^{2+}$. This substance is a deeply colored complex ion. Other complex ions that you will encounter are $Ni(NH_3)_6^{2+}$ and $CoCl_4^{2-}$. These substances result from the reaction between Ni^{2+} and NH_3 and from the reaction between Co^{2+} and Cl^-, respectively. You will also study an equilibrium involving methyl orange, an indicator. Indicators are discussed in Appendix D. Read the second, third, and fourth paragraphs of that appendix to gain an understanding of the physical and chemical properties of this indicator.

Concept of the Experiment

Le Châtelier's principle can be described in the following way: "When a system in chemical equilibrium is disturbed by a change of temperature, pressure, or concentration, the system shifts in equilibrium composition in a way that tends to counteract this change of variable" (Ebbing/Gammon, Section 14.7). This statement explains the effects that you will encounter in this experiment.

Procedure

Getting Started

1. Obtain 3 small test tubes and a piece of filter paper.
2. Obtain directions for discarding the solutions that you will use in this experiment from your laboratory instructor.
3. Be careful handling the solutions used in this experiment.

CAUTION: Solutions of ammonia, hydrochloric acid, and sodium hydroxide can cause chemical burns, in addition to ruining your clothing. Do not use your finger as a stopper when mixing these solutions. If you spill any of these solutions on you, wash the contaminated area thoroughly and immediately report the incident to your laboratory instructor. You may require further treatment.

Studying the Equilibrium of Fe^{3+} and SCN^- with $Fe(SCN)^{2+}$

1. Mark each of the test tubes with an identification number (1, 2, and 3).
2. Add 20 mL of distilled water from a graduated cylinder to a 100-mL beaker. Next add 20 drops of 0.1 *M* $Fe(NO_3)_3$ and 20 drops of 0.1 *M* KSCN to the beaker. The color is due to the $Fe(SCN)^{2+}$ ion. Stir the solution thoroughly.
3. Using a 10-mL graduated cylinder, add 3 mL of this solution to each of the test tubes.
4. Add 20 drops of 0.1 *M* $Fe(NO_3)_3$ to test tube 1. Mix the solution by shaking the tube gently.
5. Add 20 drops of 0.1 *M* KSCN to test tube 2. Mix the solution by gently shaking the tube.
6. Add 20 drops of distilled water to test tube 3 and mix. The color of the contents of this tube will serve as your reference.
7. Compare the colors in test tubes 1 and 2 with the color in the reference test tube 3. The intensity of the color in each test tube will indicate the relative concentration of $Fe(SCN)^{2+}$ in that test tube. For best results, view the test tubes down their lengths against a white paper. Record your observations.

Studying the Equilibrium of Ni^{2+} and NH_3 with $Ni(NH_3)_6^{2+}$

1. Add 10 drops of 0.1 *M* $Ni(NO_3)_2$ to a clean test tube. Record the color.
2. Add drops of 6 *M* NH_3 until the color changes and intensifies. Record the color.
3. Add drops of 6 *M* HCl until the color changes once again. Record the color. The acid has reacted with NH_3 to form NH_4^+ ions.

Studying the Equilibrium Involving Methyl Orange

1. Mark each of two small beakers with an identification letter (A for acid and B for base).
2. Add 10 mL of distilled water and 4 drops of 6 *M* HCl to the beaker marked A. Swirl the beaker to evenly distribute the acid in the water.
3. Add 10 mL of distilled water and 4 drops of 6 *M* NaOH to the beaker marked B. Swirl the beaker to evenly distribute the base in the water.
4. Add 1 mL of distilled water to a clean test tube. Then add 4 drops of the indicator solution and 2 drops of the dilute acid solution. Shake the tube gently. Record the color of the solution.
5. Add drops of the dilute base solution until the color changes. Shake the tube gently. Record the color of the solution.
6. Add drops of the dilute acid solution until the color changes again. Shake the tube gently. Record the color of the solution.

Studying the Solubility of $Ca(OH)_2$

1. Using a 10-mL graduated cylinder, add 5 mL of 6 *M* NaOH to a small, clean beaker.
2. Rinse the graduated cylinder, and then use it to add 5 mL of 1 *M* $Ca(NO_3)_2$ to the same beaker.
3. Stir the mixture thoroughly with a stirring rod. A white precipitate of $Ca(OH)_2$ should be present.
4. Using gravity filtration (described in the Introduction section of this manual), filter the mixture. This filtration may require a rather long time. While you are waiting, you may wish to begin your study of the equilibrium involving $CoCl_4^{2-}$

5. Wash the precipitate on the filter paper with 5 mL of distilled water.

6. With a metal spatula, remove as much of the wet precipitate from the filter paper as you can.

   Place the precipitate in a small, clean beaker. Add 10 mL of distilled water to the beaker. Stir the mixture to evenly distribute the solid in the water.

7. Add 2 mL of 6 *M* HCl and stir the contents of the beaker thoroughly. Record the results.

8. Add 5 mL of 6 *M* NaOH to the beaker. Record the results. You should be able to deduce the identity of the substance that is formed.

Studying the Equilibrium of Co^{2+} and Cl^- with $CoCl_4^{2-}$

1. Set up a ring stand with an iron ring. Place a piece of wire gauze on the ring. Adjust the height of the ring so that the wire gauze will be in the hottest part of the flame from a laboratory burner. Do not light the burner until this adjustment has been made.

   CAUTION: Avoid burning your fingers. Do not touch the iron ring or the wire gauze at any time while the flame is being used.

2. Place a small beaker containing distilled water on the wire gauze, and heat the water to a *gentle* boil.

3. Add 5 drops of 0.1 *M* $Co(NO_3)_2$ to a clean test tube. Record the color.

4. Add 5 drops of concentrated HCl (*do not use 6 M HCl*). Shake the test tube gently, and record the color. This is the characteristic color of the $CoCl_4^{2-}$ ion.

5. Add 5 drops of distilled water. Shake gently. Record the color.

6. Place the test tube in the boiling water, and wait a few minutes until the color has changed again. Record the color.

7. Cool the test tube in cold water or ice until the color changes once more. Record the color.

   CAUTION: Make sure that your gas outlet and those of your neighbors are closed before you leave the laboratory.

Date ______________________ Student Name ______________________

Course/Section ______________________ Team Members ______________________

Instructor ______________________ ______________________

Le Châtelier's Principle

Prelaboratory Assignment

1. Define the following terms:

   a. Chemical equilibrium

   b. Le Châtelier's principle

2. Consider the hypothetical reaction

$$A + B + \text{heat} \rightleftharpoons C + D$$

What will happen to the concentrations of A, B, C, and D under each of the following conditions?

a. A catalyst is added to the system, which is at equilibrium.

b. Either C or D is added to the system, which is initially at equilibrium.

c. Either C or D is removed from the system, which is initially at equilibrium.

d. Either A or B is added to the system, which is initially at equilibrium.

e. The system, which is initially at equilibrium, is cooled.

f. The system, which is initially at equilibrium, is heated.

3. Consider the equilibrium

$$H_2O(l) + HC_2H_3O_2(aq) \rightleftharpoons H_3O^+(aq) + C_2H_3O_2^-(aq)$$

Why will the addition of NaOH to a solution of acetic acid cause the concentration of the acetate ion ($C_2H_3O_2^-$) to increase?

Student name: _______________ Course/Section: ________________ Date: ______________

4. Write balanced chemical equations that describe the equilibria that you will observe during this experiment.

5. What special safety precautions are cited in this experiment?

Date ______________________ Student Name ______________________
Course/Section ______________________ Team Members ______________________
Instructor ______________________ ______________________

Le Châtelier's Principle

Results

1. *The equilibrium of Fe^{3+} and SCN^- with $Fe(SCN)^{2+}$*

   Compare the colors in the following pairs of test tubes.

   1 and 3: ____________

   2 and 3: ____________

2. *The equilibrium of Ni^{2+} and NH_3 with $Ni(NH_3)_6^{2+}$*

   Color before addition of NH_3: ________________

   Color after addition of NH_3: ________________

   Color after addition of HCl: ________________

3. *The equilibrium involving methyl orange*

   Color before addition of dilute HCl: ________________

   Color after addition of dilute NaOH: ________________

   Color after addition of dilute HCl: ________________

4. *The solubility of $Ca(OH)_2$*

   Give the result obtained when HCl is added to a suspension of $Ca(OH)_2$ in water.

   Give the result obtained when NaOH is added. Give the identity of the substance that is formed.

5. *The equilibrium of Co^{2+} and Cl^{-} with $CoCl_4^{2-}$*

Initial color: ____________________

Color after addition of HCl: ____________________

Color after addition of H_2O: ____________________

Color after heating: ____________________

Color after cooling: ____________________

Questions

1. Use Le Châtelier's principle to explain the different colors found in the following equilibria. Write balanced chemical equations for all chemical reactions.

   a. Fe^{3+}, SCN^{-}, and $Fe(SCN)^{2+}$

   b. Ni^{2+}, NH_3, and $Ni(NH_3)_6^{2+}$

Student name: ______________ Course/Section: ______________ Date: ______________

c. Methyl orange

2. a. How does Le Châtelier's principle explain the result you obtained when you added HCl to a suspension of $Ca(OH)_2$ in water?

b. How does Le Châtelier's principle explain the result you obtained after the addition of NaOH?

3. a. Why did adding water to the equilibrium involving $CoCl_4^{2-}$ cause the color to change? Think carefully.

b. The formation of $CoCl_4^{2-}$ from Co^{2+} and Cl^- is endothermic. Are the color changes that accompany heating and cooling of the equilibrium mixture in accord with Le Châtelier's principle? Explain.

14B. Determining an Equilibrium Constant

Introduction

Equilibria in gaseous reactions, homogeneous equilibria for substances in solution, and heterogeneous equilibria between solids and substances in solution occur frequently. All forms of homogeneous and heterogeneous equilibria are important in the laboratory and in industrial, geological, agricultural, and biological chemistries. Examples of some of these equilibria appear in your textbook. You will find that the principles of equilibria (Ebbing/Gammon, Chapter 14) are applicable to all homogeneous and heterogeneous equilibria.

Purpose

This experiment will give you an opportunity to determine the equilibrium constant for the formation of $Fe(SCN)^{2+}$. The experiment will require you to use Le Châtelier's principle.

Concept of the Experiment

When the reaction between Fe^{3+} and SCN^{-} (thiocyanate) ions in aqueous solution comes to equilibrium, the system consists of the reactants and $Fe(SCN)^{2+}$. The chemical equation for this reaction is

$$Fe^{3+}(aq) + SCN^{-}(aq) \rightleftharpoons Fe(SCN)^{2+}(aq)$$

The product is a complex ion that has a coordinate covalent bond between the iron atom and an atom (probably the S atom) from the thiocyanate anion. The color of this complex ion is so intense that thiocyanate ions can be used to detect very small quantities of Fe^{3+}. Interestingly, $Fe(SCN)^{2+}$ appears to exist solely in solution. Solid compounds containing this ion have never been isolated.

The objective of this experiment is to determine the equilibrium constant for this reaction. The equilibrium constant is given by the expression

$$K = \frac{[Fe(SCN)^{2+}]}{[Fe^{3+}][SCN]^{-}}$$

where the concentrations of the reactants (denominator) and product (numerator) are those at equilibrium. If these concentrations are measured directly or inferred from other concentrations in the solution, K can be calculated easily.

The reactants are essentially colorless and the complex ion is deeply colored. You will use a spectrophotometer (Appendix C) to monitor the absorbance due to the complex ion without interference from the reactants. The absorbance (A) is proportional to the concentration (c) of the species that absorbs the light—in this case, $Fe(SCN)^{2+}$—according to Beer's law, $A = kc$. Beer's law and the usual method for the determination of k (not to be confused with the equilibrium constant K) are discussed in Appendix C. If you did the experiment "The Absorption Spectrum of Cobalt(II) Chloride," you already know a great deal about using this law. After the concentration of $Fe(SCN)^{2+}$ has been measured by way of the absorbance, the concentrations of the reactants can be calculated from their starting concentrations and the concentration of the complex ion.

There is a problem, however. To determine k, we must measure the absorbances of a series of solutions with known amounts of the complex ion. How can known amounts of $Fe(SCN)^{2+}$ be obtained? After all, this substance is an active participant in the equilibrium with Fe^{3+} and SCN^{-} ions. Stoichiometric analysis of the quantities of the reactants will not yield a known amount of the product.

This difficulty can be avoided. Le Châtelier's principle (Ebbing/Gammon, Section 14.7) indicates that a net reaction from left to right (that is, in the forward direction) can be achieved when more of a reactant is added. As more and more of the same reactant is added, more and more of the product forms. It is possible to add so much of this reactant that essentially all of the other reactant is converted to the product. You will use limiting quantities of SCN^- and overwhelming amounts of Fe^{3+} to achieve this result. The amount of $Fe(SCN)^{2+}$ that is formed will then be essentially identical to the starting amount of the limiting reactant.

Using the technique described above, you will prepare a series of solutions with known concentrations of $Fe(SCN)^{2+}$. You will measure the absorbances of these solutions at 450 nm, the wavelength of maximum absorbance. When these absorbances are plotted against the concentrations of $Fe(SCN)^{2+}$ on a graph, k can be determined from the slope of the straight line (see Appendix C) using linear regression.

Once k has been determined, working under these conditions will no longer be advantageous. In fact, you will determine the equilibrium constant K under conditions in which substantial amounts of both of the reactants and the product are present.

Doing the Calculation

You will have to account for every dilution in order to do the calculations in this experiment. Consider, for example, the dilutions of the solution of $Fe(NO_3)_3$ that occur during the determination of k. First, 4.0 mL of a 0.0025 *M* solution of this substance is diluted to 100 mL. Portions of this solution (1.0 mL to 5.0 mL) are then diluted to 10.0 mL. The concentrations of Fe^{3+} that result are the ones to use in determining k. Do not use the original concentration. To calculate the final concentrations of the diluted solutions, you will use the familiar equation " $M_1V_1=M_2V_2$" where M_1 is the initial concentration, V_1 is the initial volume, M_2 is the concentration of the diluted solution and V_2 is the total volume of the diluted solution.

In the determination of k, you will need to construct a graph in which absorbance appears on the vertical axis and the concentration of $Fe(SCN)^{2+}$, in moles per liter, appears on the horizontal axis. Use Figure C.2 in Appendix C as an exact model.

Procedure

Getting Started

1. Your laboratory instructor may ask you to work in a group rather than alone. Be sure to wear approved goggles.
2. Obtain 5 large test tubes and 5 matching rubber stoppers. Wash, rinse, and dry the test tubes and the stoppers.
3. Mark each of the test tubes with an identification number (1 through 5).
4. Obtain directions for discarding the solutions you will use in this experiment from your laboratory instructor.
5. Obtain instructions for using your spectrophotometer from your laboratory instructor.

Determining k in Beer's Law

1. Use a Mohr pipet to transfer exactly 4.00 mL of 0.0025 *M* KSCN to a 100.0 mL volumetric flask. If such a flask is not available, use a 100-mL graduated cylinder. Add distilled water until the bottom of the meniscus coincides with the 100-mL mark on the flask or graduated cylinder. (See Figure I.2 in the Introduction section of this manual.) Add the last 0.5 mL from a medicine dropper to make sure that you do not add too much water. Mix the solution thoroughly.
2. Rinse the Mohr pipet several times with this solution. Discard each of these portions.
3. Using this pipet, the KSCN solution you have just prepared in step 1, and the volumes given in Table 14B.1, add the specified amount of this solution to each of the numbered test tubes.

Table 14B.1

Composition of Solutions for Determining *k*

Test Tube No.	Diluted KSCN (mL)	0.25 *M* $Fe(NO_3)_3$ (mL)	0.1 *M* HNO_3 (mL)
1	1.0	5.0	4.0
2	2.0	5.0	3.0
3	3.0	5.0	2.0
4	4.0	5.0	1.0
5	5.0	5.0	0

4. Pour a small amount of the 0.25 *M* $Fe(NO_3)_3$ solution into beaker. Using the iron nitrate from the beaker, rinse the pipet with 0.25 *M* $Fe(NO_3)_3$ solution. Do not put the pipet directly into the stock bottle.
5. Add the correct amount of the $Fe(NO_3)_3$ solution, as shown in the table, to each test tube.
6. Pour a small amount of 0.1 *M* HNO_3 solution into a beaker. Rinse the pipet with the 0.1 *M* HNO_3 solution. Do not put the pipet directly into the stock bottle.
7. Add the correct amount of the 0.1 *M* HNO_3 solution, as shown in the table, to each test tube. The volumes of the solutions in the test tubes should now be identical.
8. Insert the rubber stoppers. Mix each test tube thoroughly.
9. Measure and record the absorbance of each solution at 450 nm.

Determining the Equilibrium Constant

1. Wash and dry the test tubes and rubber stoppers. Renumber the test tubes (6 through 10).
2. Do not use the diluted solution of $Fe(NO_3)_3$ in this part of the experiment. Use the 0.0025 *M* solution of this substance instead.
3. Prepare the first five solutions shown in Table 14B.2. Use a properly rinsed Mohr pipet for each addition. After you mix these solutions thoroughly, measure their absorbances at 450 nm, and record your results.

Table 14B.2
Composition of Solutions for Determining *K*

Test Tube No.	0.0025 *M* $Fe(NO_3)_3$ (mL)	0.0025 *M* KSCN (mL)	0.1 *M* HNO_3 (mL)
6	1.0	1.0	5.0
7	1.0	1.5	4.5
8	1.0	2.0	4.0
9	1.0	2.5	3.5
10	1.0	3.0	3.0
11	2.0	1.0	4.0
12	2.0	1.5	3.5
13	2.0	2.0	3.0
14	2.0	2.5	2.5
15	2.0	3.0	2.0

4. Wash and dry the test tubes. Renumber them (11 through 15), and prepare the remaining solutions. Measure the absorbances of these solutions at the same wavelength, and record your results.

Date		Student Name	
Course/Section		Team Members	
Instructor			

Determining an Equilibrium Constant

Prelaboratory Assignment

1. Provide definitions for the following terms:
   a. Chemical equilibrium

   b. Homogeneous equilibrium

   c. Heterogeneous equilibrium

   d. Le Châtelier's principle

   e. Complex ion

f. Beer's law

2. a. What reaction are you studying in this experiment?

b. Give the mathematical expression for the equilibrium constant that pertains to this reaction.

c. What is the difference between k and K ?

d. How will Le Châtelier's principle be used to obtain k?

Date		Student Name	
Course/Section		Team Members	
Instructor			

Determining an Equilibrium Constant

Results

1. *Obtaining data for determining k*

Test Tube No.	*A*
1	
2	
3	
4	
5	

2. *Obtaining data for determining K*

Test Tube No.	*A*	Test Tube No.	*A*
6		11	
7		12	
8		13	
9		14	
10		15	

Questions

1. a. Complete the following table by calculating $[Fe(SCN)^{2+}]$ for each test tube and inserting the absorbance as a first step in determining *k*. All concentrations are to be given in *M*.

Test Tube No.	$[Fe(SCN)^{2+}]$	A
1	______________	______________
2	______________	______________
3	______________	______________
4	______________	______________
5	______________	______________

Representative calculation:

b. Use the available piece of graph paper to plot these absorbances and values of $[Fe(SCN)^{2+}]$.

Student name: ________________ Course/Section: __________________ Date: _______________

c. Calculate the slope of the best straight line, using linear regression (see Appendix C). Either do the calculation by hand or, if your laboratory instructor wishes, use the tool available online at the student website. If you wish, draw a straight line with this slope on your graph. The line should pass through the origin ($A = 0$, $[Fe(SCN)^{2+}] = 0$).

k = ____________________(Give units.)

2. a. Complete the following table (with all concentrations in *M*).

Test Tube No.	Starting $[Fe^{3+}]$	Starting $[SCN^-]$	$[Fe(SCN)^{2+}]$	Equilibrium $[Fe^{3+}]$	Equilibrium $[SCN^-]$	*K*
6						
7						
8						
9						
10						
11						
12						
13						
14						
15						

Mean: ____________

Representative calculation:

Student name: ________________ Course/Section: __________________ Date: _______________

b. Determine the precision of your results for K, using the method shown in Appendix A. Do the calculation by hand, or, if your laboratory instructor wishes, use the tool available online at the student website.

3. Although the use of absorbances at 450 nm provided you with maximum sensitivity, the absorbances at, say, 400 nm or 500 nm are not zero and could have been used throughout this experiment. Would the same value of K be obtained at one of these wavelengths? Explain.

15. The Relative Strengths of Some Acids

Introduction

Acids and bases, two closely connected types of substances, are part of our everyday lives. They can be found in foods, soft drinks, medicines, and cleaning products.

The relationship between acids and bases occurs because these substances react readily with each other. An acid–base reaction is a competition for protons (Ebbing/Gammon, Chapter 15). The extent of the reaction depends on the strengths of the acid and the base. A strong acid reacts with a base more completely than a weaker acid reacts with the same base. This difference enables us to measure the relative strengths of acids (and bases).

Purpose

You will estimate or measure the pH of various solutions, using indicators, pH paper, and, if available, a pH meter.

Relative Strengths of Acids

How can you determine the relative strengths of two acids? As an example, let us consider how you would determine the relative strengths of HF (hydrofluoric acid) and HCN (hydrocyanic acid). You would prepare *isomolar* aqueous solutions of these acids and examine the extent of their ionizations in water. Isomolar solutions have the same formal molarity but may differ in the extent of ionization. The equations that describe the ionization reactions for these two acids are

$$HF(aq) + H_2O(l) \rightleftharpoons F^-(aq) + H_3O^+(aq)$$

$$HCN(aq) + H_2O(l) \rightleftharpoons CN^-(aq) + H_3O^+(aq)$$

The ionization of the stronger acid will be greater than the ionization of the weaker acid. As a result, the stronger acid will provide more H_3O^+ ions than the weaker acid. You would find in this case that $[H_3O^+]$ is greater in the solution of HF. You would then know that HF is a stronger acid than HCN.

The relative strengths of the acids in Table 15.1 were determined by this method. The table also shows the relative strengths of the conjugate bases of the acids. When you compare the strengths of any two acids in the table, you will note that the stronger acid has the weaker conjugate base.

Two Special Notes

The aluminum cation, Al^{3+}, exists in aqueous solution as the hydrated ion, $Al(H_2O)_6^{3+}$. Table 15.1 shows that this ion is an acid. Why? The water molecules are bonded to the metal through lone pairs of electrons on the oxygen atoms. The positively charged aluminum ion draws electrons from the oxygen atoms, which, in turn, draw electrons from the O–H bonds and weakens them. As a result, the water molecules tend to be acidic:

$$Al(H_2O)_6^{3+}(aq) + H_2O(l) \rightleftharpoons Al(H_2O)_5OH^{2+}(aq) + H_3O^+(aq)$$

Table 15.1

Relative Strengths of Some Acids and Bases

	Acid	Base	
Strongest acids	$HClO_4$	ClO_4^-	Weakest bases
↓	H_2SO	HSO_4^-	↑
	HI	I^-	
	HBr	Br^-	
	HCl	Cl^-	
	HNO_3	NO_3^-	
	H_3O^+	H_2O	
	HSO_4^-	SO_4^{2-}	
	H_2SO_3	HSO_3^-	
	H_3PO_4	$H_2PO_4^-$	
	HNO_2	NO_2^-	
	HF	F^-	
	$HC_2H_3O_2$	$C_2H_3O_2^-$	
	$Al(H_2O)_6^{3+}$	$Al(H_2O)_5OH^{2+}$	
	H_2CO_3	HCO_3^-	
	H_2S	HS^-	
	$HClO$	ClO^-	
	$HBrO$	BrO^-	
	NH_4^+	NH_3	
	HCN	CN^-	
	HCO_3^-	CO_3^{2-}	
	H_2O_2	HO_2^-	
	HS^-	S^{2-}	
Weakest acids	H_2O	OH^-	Strongest bases
	NH_3	NH_2^-	
	OH^-	O^{2-}	

You will also note that Table 15.1 indicates that the ammonium ion, NH_4^+, is an acid. The acidity of this ion is a result of the reaction shown in the following chemical equation:

$$NH_4^+(aq) + H_2O(l) \rightleftharpoons NH_3 + H_3O^+(aq)$$

pH

The concentration of H_3O^+ ions in a solution is usually determined by measuring the pH of the solution. The pH is defined as

$$pH = -\log[H_3O^+]$$

(Ebbing/Gammon, Section 15.8). If the pH of the solution is low, the ionization equilibrium favors the right side of the chemical equation because $[H_3O^+]$ is large. However, if the pH is high, the equilibrium favors the left side because $[H_3O^+]$ is small. When two acids are compared, the stronger acid yields a solution with the lower pH.

Concept of the Experiment

You will estimate or measure the pH of isomolar solutions of HCl, H_3PO_4, $HC_2H_3O_2$, NaH_2PO_4, $Al(NO_3)_3$, $Zn(NO_3)_2$, and NH_4NO_3. Many of these substances can be found in Table 15.1. You will be able to judge the strengths of these acids from the pH values of their solutions. You will also have an opportunity to estimate or measure the pH of several common substances: vinegar, carbonated water, tap water, and distilled water

Procedure

Getting Started

1. Obtain 4 large test tubes. Use a graduated cylinder to place 4 mL of distilled water in each test tube. Mark the height of the meniscus in each tube with a marking pencil. Pour out the water and dry the test tubes.
2. Obtain a small (about 1/2-inch) strip of wide-range pH paper for each solution that you will test.

   Handle these pieces as little as possible as chemicals from your hands may react with the acid/base indicators coating the paper. Place the strips of pH paper on a clean, dry paper towel.
3. Obtain a small (about 1/2-inch) strip of each narrow-range pH paper that is available. Handle these pieces as little as possible. Place them on a clean, dry paper towel. Mark the pH range for each strip on the paper towel. Replace these strips as needed.
4. If you are going to use a pH meter, obtain directions for using it from your laboratory instructor.

Doing the Experiment with Indicators and pH Paper

1. Complete Steps 2 through 6 for each solution whose pH is to be estimated.
2. Make sure your test tubes are clean and dry.
3. Obtain 4 mL of the solution* to be tested in each of the test tubes.
4. Using a clean, dry stirring rod, place 1 drop of the solution from one of the test tubes on a piece of wide-range pH paper. Do not contaminate the other unused strips. Using the result that you obtain, select the appropriate narrow-range paper. Place 1 drop of the solution on the narrow-range paper. Record your estimate of the pH of the solution.
5. Use 3 drops of a different indicator solution for each of the four test tubes. The indicators are thymol blue, methyl orange, methyl red, and bromthymol blue (see Figure D.1 in Appendix D for the pH range for these indicators).
6. Estimate the pH by comparing your results with Figure D.1. In some instances, you may be able to estimate to the nearest 0.1 pH unit; in other instances, you may be able to estimate only to the nearest 0.5 pH unit. Record your estimate of the pH of the solution.

   **Note:* Instructors, please refer to the IRM for acceptable concentration range for these solutions.

Measuring the pH with a pH Meter (optional)

1. Calibrate the pH meter with solutions of known pH according to the methods described by your laboratory instructor.
2. Measure, and record to the nearest 0.01 pH unit, the pH of each solution.

Date ______________________ Student Name ______________________

Course/Section ______________________ Team Members ______________________

Instructor ______________________ ______________________

The Relative Strengths of Some Acids

Prelaboratory Assignment

1. Provide definitions for the following terms:

   a. Brønsted–Lowry acid

   b. Brønsted–Lowry base

   c. pH

   d. Acidic solution

   e. Basic solution

   f. Neutral solution

   g. Indicator

   h. pH paper

   i. pH meter

2. a. A solution causes a yellow color with thymol blue, an orange color with methyl orange, a red color with methyl red, and a yellow color with bromthymol blue. The pH must lie between _____________ and _____________. The estimated pH is _____________ ± _____________.

b. A solution causes a yellow color with thymol blue, yellow colors with both methyl orange and methyl red, and a green color with bromthymol blue. The pH must lie between _____________ and _____________. The estimated pH is _____________ ± _____________.

ate		Student Name	
ourse/Section		Team Members	
ıstructor			

The Relative Strengths of Some Acids

Results

Solution	pH from pH Paper	pH from Indicators	pH from pH Meter

Questions

1. a. Calculate the concentration of H_3O^+ ions in the most acidic solution that you examined. Use your most precise pH measurement.

b. Calculate the concentration of H_3O^+ ions in the least acidic solution that you examined. Use your most precise pH measurement.

2. a. Arrange the acids that you used in order of *decreasing* acidity. Use your most precise pH measurements. Exclude the common household substances.

   b. How does your arrangement compare with the first column in Table 15.1, insofar as a comparison can be made? Comment on the similarities and differences.

   c. Comment on the relative acidities of H_3PO_4 and the $H_2PO_4^+$ ion.

3. Write the balanced chemical reactions that show why solutions of $Al(NO_3)_3$, $Zn(NO_3)_2$, and NH_4NO_3 have the pH values that you found.

4. Arrange the common household substances in order of *decreasing* acidity.

16A. Equilibria with Weak Acids and Weak Bases

Introduction

Equilibria with weak acids and weak bases are subjects of considerable importance (Ebbing/Gammon, Chapter 16). Your study of these equilibria will begin with solutions containing a single solute. The solute will be either a weak acid or a weak base. Salts that *hydrolyze* will be included in this category. When a salt hydrolyzes, it reacts with water to form its conjugate acid and hydroxide ion or its conjugate base and hydronium ion. In the next category, you will consider solutions with two solutes. Using two solutes will allow you to study the common-ion effect and behavior of buffer solutions.

Purpose

This experiment will allow you to examine the effect of dilution on the degree of ionization of a weak acid and a weak base, the pH of a solution containing a polyprotic acid, the change in the pH of a solution resulting from the common-ion effect, and the pH behavior of buffer solutions.

Concept of the Experiment

Each part of this experiment and your interpretation of the results depend on your estimate or measurement of pH. You will use either pH paper or a pH meter. These methods, which may be familiar to you from the experiment "The Relative Strengths of Some Acids," are discussed in Appendix D.

Procedure

Getting Started

1. Your laboratory instructor may ask you to work with a partner.
2. If you are using a pH meter for the first time, obtain directions on its proper use from your laboratory instructor.

Table 16A.1

Composition of Solutions

Solution	Composition
1	0.10 $M\ HC_2H_3O_2$
2	5 mL 0.10 $M\ HC_2H_3O_2$ + 5 mL H_2O
3	1 mL 0.10 $M\ HC_2H_3O_2$ + 99 mL H_2O
4	5 mL 0.10 $M\ HC_2H_3O_2$ + 5 mL 0.10 M HCl
5	0.10 $M\ H_3PO_4$
6	0.10 $M\ NH_3$
7	0.10 $M\ NH_4NO_3$
8	50 mL 0.10 $M\ NH_3$ + 50 mL 0.10 $M\ NH_4NO_3$
9	10 mL Solution 8 + 6 mL H_2O
10	10 mL Solution 8 + 5 mL H_2O + 1 mL 0.10 M HCl
11	10 mL Solution 8 + 6 mL 0.10 M HCl
12	10 mL Solution 8 + 5 mL H_2O + 1 mL 0.10 M NaOH
13	10 mL 0.10 $M\ HC_2H_3O_2$ + 5 mL 0.10 M NaOH
14	10 mL 0.10 $M\ NH_4NO_3$ + 5 mL 0.10 M NaOH

Doing the Experiment

1. Prepare the solutions in Table 16A.1 one at a time in clean, dry glassware. Always use distilled water.
2. After you prepare each of the solutions, mix it thoroughly. If you use a stirring rod, make sure it is clean and dry. Estimate or measure the pH and record the result.
3. Rinse and dry the glassware before using it again.
4. Discard solutions as directed by your laboratory instructor.

Date		Student Name	
Course/Section		Team Members	
Instructor			

Equilibria with Weak Acids and Weak Bases

Prelaboratory Assignment

1. Define the following terms:

   a. Acid-ionization constant (K_a)

   b. Base-ionization constant (K_b)

   c. Degree of ionization

   d. Percent ionization

   e. Common-ion effect

   f. Buffer

2. a. Which substances in this experiment are strong acids and strong bases?

b. Which are weak acids and weak bases?

c. Which solutions in Table 16A.1 should exhibit common-ion effects?

d. Which solutions are buffers?

Date ____________________ Student Name ____________________
Course/Section ____________________ Team Members ____________________
Instructor ____________________ ____________________

Equilibria with Weak Acids and Weak Bases

Results

Solution No.	pH	Solution No.	pH
1	________	8	________
2	________	9	________
3	________	10	________
4	________	11	________
5	________	12	________
6	________	13	________
7	________	14	________

Questions

1. a. Calculate the degree of ionization of acetic acid in Solutions 1 through 3 (Ebbing/Gammon, Section 16.1).

   b. How do your results compare with the expected behavior?

Student name: ________________ Course/Section: ________________ Date: ________________

2. Calculate the expected pH of the following solutions (Ebbing/Gammon, Sections 16.1, 16.2, and 16.3). Compare the calculated values with your experimental results from Solutions 5, 6, and 7.

   a. 0.10 M H_3PO_4 with $K_{a1} = 6.9 \times 10^{-3}$:

   b. 0.10 M NH_3 with $K_b = 1.8 \times 10^{-5}$:

   c. 0.10 M NH_4NO_3:

3. a. Compare Solutions 2 and 4. Explain how the common-ion effect influences the pH of Solution 4? Calculate the expected pH of each of these solutions.

b. Explain how the common-ion effect influences the pH of Solution 8? Compare the observed pH with the calculated value.

Student name: ______________ Course/Section: ________________ Date: ______________

4. a. How do Solutions 8, 10, 11, and 12 show the properties of a buffer solution?

b. Calculate the expected pH for each of these solutions.

c. Should the pH of a buffer change when the buffer is diluted? Explain fully, using the Henderson–Hasselbalch equation (Ebbing/Gammon, Section 16.6) as well as your results from Solutions 8 and 9.

5. What is responsible for the pH behavior of each of the following solutions? Include a chemical equation in your explanation. Calculate the pH you would expect from each solution.

   a. Solution 13

   b. Solution 14

16B. An Acid–Base Titration Curve

Introduction

Acid–base titration curves (Ebbing/Gammon, Section 16.7) are graphs that show the successive pH values that occur during the titration of a base with an acid or of an acid with a base. A typical titration curve for the titration of an acid with a base can be found in Figure 16B.1.

The general purpose of a titration is to determine the amount of a particular substance in a sample (Ebbing/Gammon, Section 4.10). An indicator is usually employed to show when a stoichiometric amount of another substance has been added from a buret. An example appears in the experiment "How Much Acetic Acid Is in Vinegar?" The general purpose can also be achieved with a titration curve, but the procedure for obtaining the required data is much slower than one employing an indicator. However, a titration curve will enable an analyst to choose an indicator for subsequent titrations of similar samples with the same reagent.

When a weak acid is titrated with a strong base, or a weak base is titrated with a strong acid, the titration curve is unique for the weak acid or the weak base. As a consequence, there is another use for a titration curve: It can be used to determine the ionization constants for weak acids and weak bases.

FIGURE 16B.1

An acid–base titration curve resulting from the titration of a solution of acetic acid with a 0.101 *M* solution of NaOH. The equivalence point occurs after the addition of 27.02 mL of the NaOH solution.

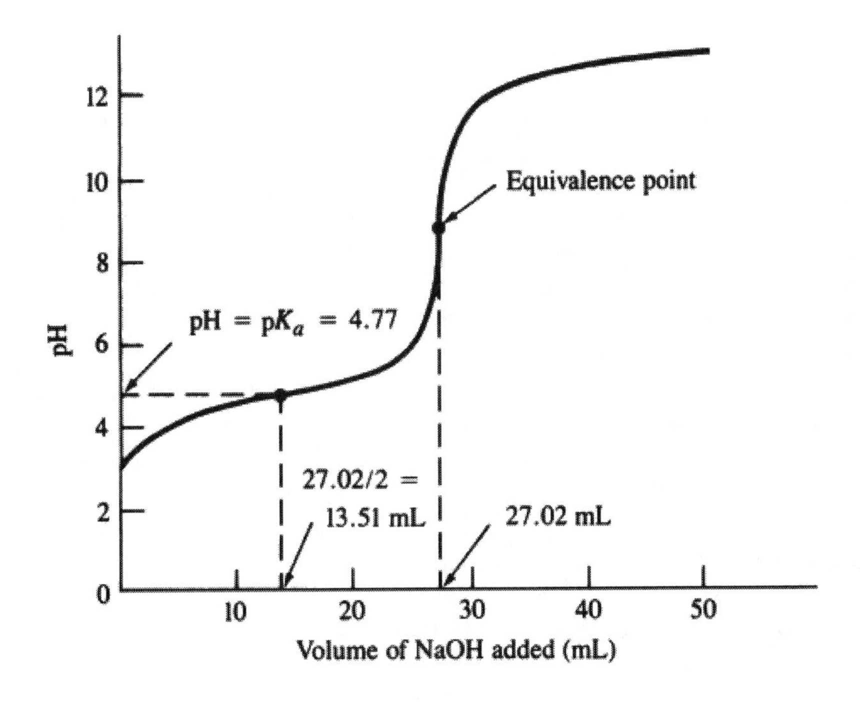

Purpose

You will construct an acid–base titration curve on a piece of graph paper. You will gather the data that you need for this graph by titrating a weak acid, potassium hydrogen phthalate, with a solution of sodium hydroxide. The titration curve will be used to achieve two goals: standardizing the solution of sodium hydroxide, and illustrating the method for obtaining the acid-ionization constant for the weak acid.

Standardization of a NaOH Solution

You cannot prepare a solution of sodium hydroxide with an accurately known concentration by adding a known mass of NaOH to an appropriate quantity of water. The *hydroscopic* (readily taking and retaining water from air) nature of solid NaOH makes measuring the mass of a sample of this substance with accuracy virtually impossible. A significant fraction of the measured mass would be due to an unknown quantity of water adsorbed from the air.

Another approach is available, however; the reaction of NaOH with a known quantity of an acid can be used to determine the molarity of the NaOH solution. Potassium hydrogen phthalate ($KHC_8H_4O_4$, abbreviated here as KHP), a weak acid, is generally used for this purpose. The structure of KHP is shown in Figure 16B.2.

FIGURE 16B.2

The molecular structure of potassium hydrogen phthalate (KHP).

COOH

COO^-K^+

KHP

This substance is an ionic, monoprotic acid that exists in solution as K^+ cations and, primarily, $HC_8H_4O_4^-$ anions. The ionization equilibrium for this weak acid is

$$HC_8H_4O_4^-(aq) + H_2O(l) \rightleftharpoons H_3O^+(aq) + C_8H_4O_4^{2-}(aq)$$

The exact molarity of the NaOH solution is calculated (Ebbing/Gammon, Section 4.10) from the mass of KHP, its molar mass, and the volume of the NaOH solution required to react completely with the KHP, as determined by a titration. The entire procedure by which we obtain the molarity of a solution of one substance from an accurately known amount of another substance is called *standardization.*

The Acid-Ionization Constant

Take a second look at the titration curve in Figure 16B.1. A solution of acetic acid has been titrated with a 0.101 *M* solution of NaOH. The equivalence point (Ebbing/Gammon, Section 16.7) occurs after the addition of 27.02 mL of the NaOH solution. This titration curve also provides enough data to determine the ionization constant for acetic acid.

The Henderson–Hasselbalch equation (Ebbing/Gammon, Section 16.6),

$$\text{pH} = \text{pK} + \log\frac{[\text{base}]}{[\text{acid}]}$$

plays a crucial role in finding the ionization constant, K. Although this equation can be used to calculate the pH of a buffer (Ebbing/Gammon, Section 16.6), it can also be used to calculate the pH of a point on the titration curve if that point lies within a particular range. The point must lie after the initial point (where none of the NaOH solution has been added) and before the equivalence point where the number of moles of acetic acid equals the number of moles of NaOH. As you read about applying this equation, remember that K_a is the ionization constant of the weak acid (acetic acid in this example); that [acid] is the concentration of the weak acid; and that [base] is the concentration of the weak acid's conjugate base.

During the titration, [acid] will be decreasing because the weak acid is reacting with each increment of NaOH. And since the conjugate base of the weak acid is the product of this reaction, [base] will be increasing. These quantities, [acid] and [base], must become equal at some point. The equality will occur halfway to the equivalence point. At this halfway point, half of the NaOH required to react completely with the weak acid will have been added. As a result, half of the weak acid will have been converted to its conjugate base, and so the molar quantities of these substances will be identical. The halfway point in Figure 16B.1 occurs at 27.02/2, or 13.51 mL. The equality of [acid] and [base] allows us to write

$$[\text{acid}] = [\text{base}]$$

$$\log\frac{[\text{base}]}{[\text{acid}]} = \log 1 = 0$$

The Henderson–Hasselbalch equation reduces to

$$\text{pH} = \text{p}K_a$$

at the halfway point.

After 13.51 mL is located on the titration curve, the corresponding pH is read directly from the graph. This pH turns out to be 4.77, and pK_a for acetic acid must also be 4.77. The acid-ionization constant, which is calculated from pK_a, is

$$K_a = 10^{-4.77} = 1.7 \times 10^{-5}$$

This result agrees with the known value of K_a (Ebbing/Gammon, Table 16.1).

The Location of the Equivalence Point

The most accurate method for finding the equivalence point after the acid–base titration curve has been drawn involves the behavior of $\Delta(\Delta\text{pH}/\Delta V)\Delta V$. This quantity can also be written as $\Delta^2\text{pH}/\Delta V^2$. Those familiar with calculus will know that this is a second derivative. The equivalence point corresponds to the volume at which the second derivative is zero. The use of the second derivative, however, would complicate this experiment unnecessarily.

Instead, you will locate the approximate equivalence point using the following graphical procedure. A titration curve can be approximated by three straight lines, as shown in Figure 16B.3. Two of the lines intersect at A, and another intersection occurs at B. The approximate equivalence point will be located at the midpoint of the *vertical* line between A and B.

FIGURE 16B.3

A method for locating the approximate equivalence point. Note that it lies at the midpoint of the *vertical* line between A and B.

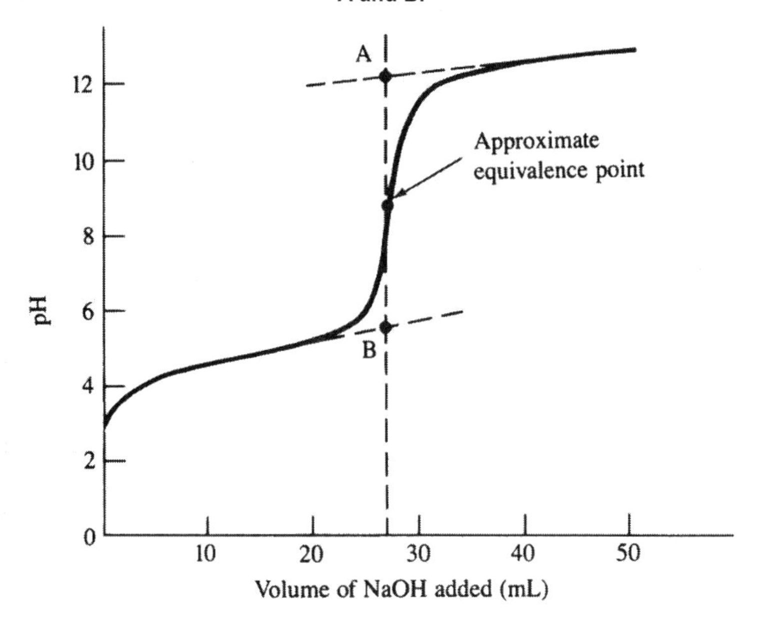

Concept of the Experiment

You will weight a sample of KHP and titrate it with a solution of NaOH. The approximate concentration of that solution will be 0.1 *M*. You willcalculate the approximate mass of KHP that you need for the titration as part of the Prelaboratory Assignment.

The NaOH solution will be delivered from a *buret*. Make sure you are familiar with the methods for cleaning, filling, and using a buret; these are described in the Introduction section of this manual. However, you will not be able to use the titration technique shown in Figure I.4 because of the presence of the electrodes. A beaker will be used instead of an Erlenmeyer flask. The solution in the beaker will have to be stirred, either by hand or mechanically, rather than swirled. Stirring is important because the pH will tend to drift until a completely homogeneous solution is achieved. General instructions for using pH meters can be found in Appendix D. Your laboratory instructor will provide you with instructions on how to properly use the pH meters found in your laboratory.

Follow all instructions carefully during the titration. Remember that the titration must provide sufficient data without being too time consuming.

Procedure

Getting Started

1. Your laboratory instructor may ask you to work with a partner.
2. Obtain a 50-mL buret.
3. If you are going to use a pH meter for the first time, obtain instructions for using it.

Cleaning and Filling Your Buret

1. Instructions for using a buret can be found in the Introduction section of this manual. Clean your buret and fill it with the NaOH solution that is available in the laboratory. This solution is approximately 0.1 *M*.

Doing the Titration

1. Measure the mass of a piece of waxed weighing paper, using your most precise balance. Record the mass.
2. Add potassium hydrogen phthalate (KHP) to the paper until you have obtained the quantity that you calculated in the Prelaboratory Assignment. Record the combined masses of the paper and the KHP. Calculate and record the mass of the KHP.
3. Carefully pour the KHP from the paper into a clean, dry 150-mL beaker. Add about 50 mL of distilled water, and swirl gently until the sample has dissolved.
4. Calibrate the pH meter following the instructions provided by your laboratory instructor. A buffer solution whose pH is 5.00 is preferred. Be sure to rinse the electrodes with distilled water and dry them gently with tissue paper before and after placing them in the buffer solution.
5. Immediately immerse the electrodes in the solution of KHP. If possible, clamp the electrodes so that they are not touching the beaker.
6. Provide some means of stirring the solution. A magnetic stirring device is preferred. However, stirring by hand with a glass stirring rod will be acceptable if you are thorough and if you avoid hitting the electrodes.
7. The buret should be clamped so that its tip is within the beaker but above the surface of the solution.
8. Record the initial buret reading. For accurate results, remember to estimate the volume to the nearest 0.01 mL.
9. Read and record the initial pH of the solution before any of the NaOH solution has been added.
10. Begin the titration by adding successive portions of about 1 mL of the NaOH solution. Obtain and record the buret reading and the pH after each addition.
11. When the pH begins to increase by more than about 0.3 pH unit after an addition, decrease the portions that you add to about 0.2 mL. Once the equivalence point has been passed, the pH change after each addition will decrease. When the change is less than or equal to 0.3 pH unit, return to 1-mL portions. Continue the titration until the pH of the solution is about 11.5–12.
12. Repeat Steps 1 through 11 with a new sample of KHP (if time permits).

Date	______________________	Student Name	______________________
Course/Section	______________________	Team Members	______________________
Instructor	______________________		______________________

An Acid–Base Titration Curve

Prelaboratory Assignment

1. Provide definitions for the following terms:

   a. Acid–base titration curve

   b. Equivalence point

   c. Standardization

   d. pH

   e. pH meter

2. Estimate the mass of KHP that will require 25 mL of 0.10 *M* NaOH to reach the equivalence point in a titration. Give the balanced chemical equation for the reaction.

3. Calculate the molarity of the acetic acid solution in Figure 16B.1 if 25.0 mL of that solution has been titrated with the 0.101 *M* solution of NaOH.

Date		Student Name	
Course/Section		Team Members	
Instructor			

An Acid–Base Titration Curve

Results

Sample	1	2
Mass of KHP and paper (g)		
Mass of weighing paper (g)		
Mass of KHP (g)		

Sample 1
Initial pH: ________________

Initial buret reading (mL): ________________

Buret Reading (mL)	Volume Added (mL)	pH	Buret Reading (mL)	Volume Added (mL)	pH

Student name: ________________ Course/Section: _________________ Date: _______________

Sample 2

Initial pH: _________________

Initial buret reading (mL): _________________

Buret Reading (mL)	Volume Added (mL)	pH	Buret Reading (mL)	Volume Added (mL)	pH

Questions

1. a. Use the graph paper that is available to plot the titration curves. Locate the equivalence point on each graph. Complete the following table.

Sample	1	2
Volume at equivalence point (mL)	________	________
pH at equivalence point	________	________

b. Calculate the molarity of the NaOH solution from each result, and calculate the mean.

c. Use your graphs to obtain the data required in the following table.

Sample	1	2
Volume at halfway point (mL)	________	________
pH at halfway point	________	________
pK_a	________	________
Mean pK_a	________	

Student name: ______________ Course/Section: ______________ Date: ______________

d. Use the mean pK_a to calculate K_a for KHP.

2. Use your value of K_a to calculate the initial pH for each of your samples (Ebbing/Gammon, Section 16.1), and compare the calculated and experimental results.

3. a. Calculate K_b for the $C_8H_4O_4^{2-}$ anion from the value of K_a that you obtained for KHP.

b. Use K_b, the number of moles of $C_8H_4O_4^{2-}$ at the equivalence point, and the total volume at that point to calculate the pH for each sample at the equivalence point. Compare these calculated results with the experimental results.

Student name: ________________ Course/Section: _________________ Date: _______________

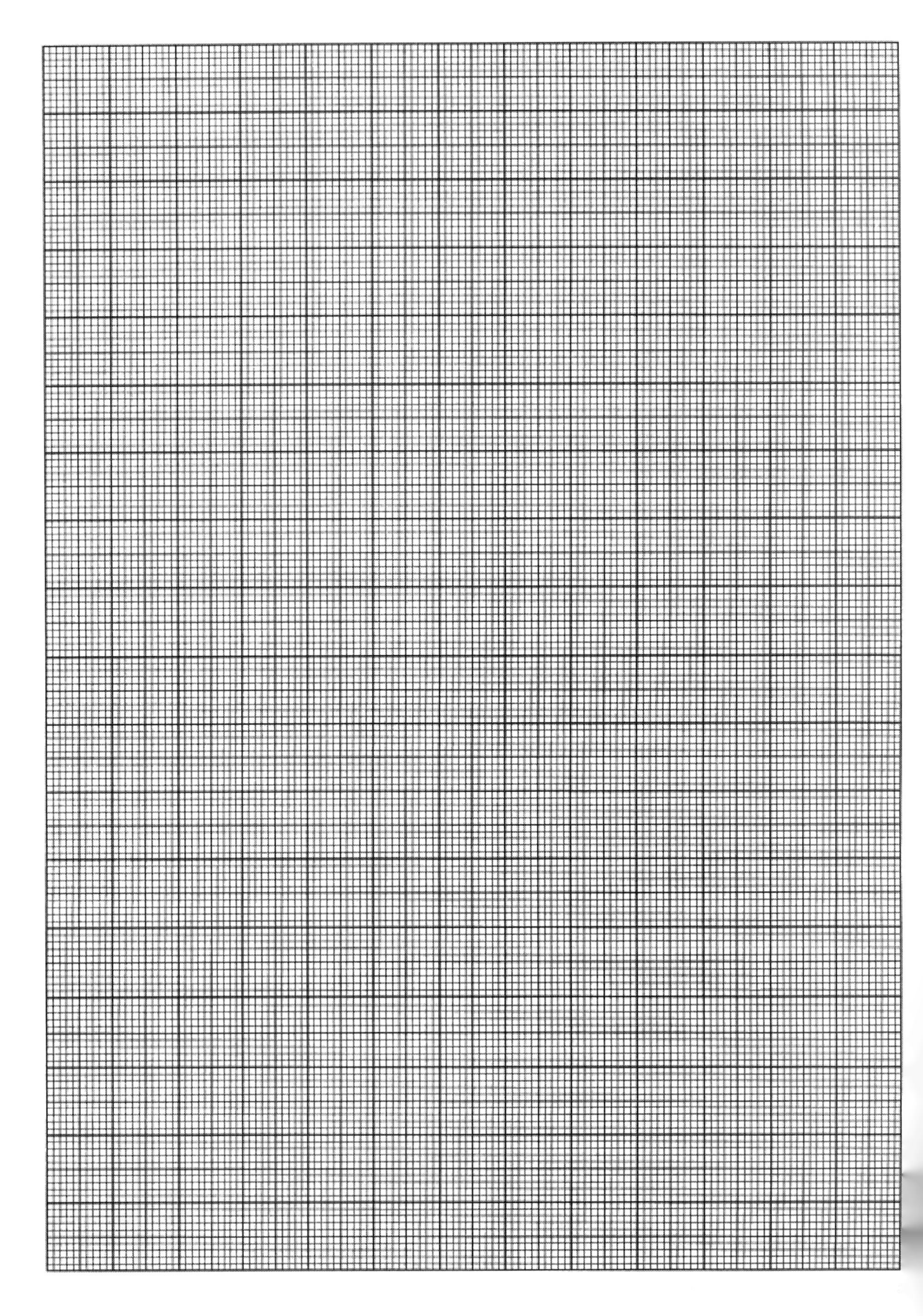

17A. A Solubility Product Constant

Introduction

The interaction of a slightly soluble ionic compound with its dissolved ions leads to another type of equilibrium (Ebbing/Gammon, Chapter 17). The equilibrium constant for this kind of equilibrium has a special name. It is called the *solubility product constant.*

Purpose

In this experiment, you will determine the solubility of $Ca(IO_3)_2$ (calcium iodate) and calculate its solubility product constant. You will also demonstrate that the solubility of $Ca(IO_3)_2$ is independent of the amount of water that is used to obtain a saturated solution of this substance.

Concept of the Experiment

If the solubility of a slightly soluble ionic compound is measured, the solubility product constant for the substance can be calculated (Ebbing/Gammon, Example 17.3). This is the method that you will adopt for $Ca(IO_3)_2$ in this experiment. The equilibrium between solid $Ca(IO_3)_2$ and its ions in a saturated solution is

$$Ca(IO_3)_2(s) \rightleftharpoons Ca^{2+}(aq) + 2IO_3^-(aq)$$

If some analytical technique is used to determine the concentration of either Ca^{2+} ions or IO_3 ions in the saturated solution, the solubility of $Ca(IO_3)_2$ will be known, and the solubility product constant (K_{sp}) can be calculated.

In this experiment, you will determine the concentration of IO_3^- ions through a titration with a standardized solution of $Na_2S_2O_3$ (sodium thiosulfate) in the presence of KI (potassium iodide), using starch as an indicator. A definition of standardization can be found in the experiment "An Acid–Base Titration Curve." Potassium iodide will react with the iodate ions to give I_2 as the sole product containing iodine. The reaction is

$$IO_3^-(aq) + 5I^-(aq) + 6H_3O^+(aq) \rightarrow 3I_2(aq) + 9H_2O(l)$$

The molecular iodine reacts with S_2O^{2-} ions during the titration according to

$$I_2(aq) + 2S_2O_3^{2-}(aq) \rightarrow 2I^-(aq) + S_4O_6^{2-}(aq)$$

where $S_4O_6^{2-}$ is the tetrathionate ion. When these two equations are combined, multiplying the second reaction by 3 so the intermediate I_2 is eliminated, the net reaction occurring in the titration is obtained.

$$\begin{array}{c} IO_3^-(aq) + 5I^-(aq) + 6H_3O^+(aq) \rightarrow 3I_2(aq) + 9H_2O(l) \\ 3I_2(aq) + 6S_2O_3^{2-}(aq) \rightarrow 6I^-(aq) + 3S_4O_6^{2-}(aq) \\ \hline IO_3^-(aq) + 6S_2O_3^{2-}(aq) + 6H_3O^+(aq) \rightarrow I^-(aq) + 3S_4O_6^{2-}(aq) + 9H_2O(l) \end{array}$$

Starch is used as an indicator in this titration because it reacts with I_2 reversibly to form a dark blue color. I_2 is consumed in the titration, so the color fades as the titration progresses. You will know when a stoichiometric volume of the $Na_2S_2O_3$ solution has been added because at that point, one drop of that solution will cause the disappearance of the last trace of the blue color. A trial titration will enable you to find the approximate volume that is required before you do the first of two exact titrations.

Your laboratory instructor may ask you to standardize the $Na_2S_2O_3$ solution, using a known volume of a KIO_3 solution whose molarity is known accurately. If you are asked to do this, a calculation in the Prelaboratory Assignment will make the task easier. Alternatively, the standardized solution of $Na_2S_2O_3$ may be provided for you.

Procedure

Getting Started

1. Obtain a 10-mL transfer pipet, a 50-mL buret, and 3 pieces of filter paper.
2. Ask your laboratory instructor whether you are to standardize the $Na_2S_2O_3$ solution.

Preparing Saturated Solutions of $Ca(IO_3)_2$

1. Prepare $Ca(IO_3)_2$ by adding 50 mL of 0.2 *M* KIO_3 to 20 mL of 1 *M* $Ca(NO_3)_2$ in a 150-mL beaker. Do not use the 0.0100 *M* solution of KIO_3 that will be available if you are to standardize your $Na_2S_2O_3$ solution.
2. Stir the mixture vigorously with a stirring rod. A white, crystalline precipitate of $Ca(IO_3)_2$ should form.
3. Let the mixture stand for a few minutes while you prepare for gravity filtration (see the Introduction section of this manual for details on the gravity filtration method).
4. Filter the precipitate. Rinse the remnants from the beaker onto the filter paper, using distilled water from a plastic wash bottle.
5. Wash the precipitate on the filter paper with three small portions of distilled water.
6. Place about 1/3 of the wet precipitate into a clean, labeled beaker, using a metal spatula. Place a similar portion of the precipitate into an identical beaker, using the same method. These beakers should have capacities of at least 100 mL. Save the remaining portion of the precipitate on the filter paper to use if an unforeseen accident occurs.
7. Use a graduated cylinder to add 40 mL of distilled water to the first beaker and 80 mL of distilled water to the second beaker.
8. Stir each beaker thoroughly, using separate stirring rods.
9. Allow each stirring rod to remain in its beaker, and let each mixture stand for at least 30 min with occasional stirring.
10. Go on to the next parts of this experiment while you are waiting for the calcium iodate to dissolve.

Cleaning and Filling Your Buret

1. The Introduction section of this manual provides instructions for using a buret. Clean your buret and fill it with the $Na_2S_2O_3$ solution, following those directions.

Standardizing the $Na_2S_2O_3$ Solution (optional)

1. Pipet 10.0 mL of 0.0100 *M* KIO_3 into each of two clean 125-mL or 250-mL Erlenmeyer flasks. Do not use the 0.2 *M* solution of KIO_3.
2. Add about 20 mL of distilled water to each flask from a clean graduated cylinder.

3. Dissolve about 1 cm^3 of solid KI, as measured in a clean, dry 10-mL graduated cylinder, in each solution. Add 20 drops of 2 *M* HCl to each flask, and swirl to obtain homogeneous solutions.

CAUTION: Handle hydrochloric acid with care. It can cause chemical burns, in addition to ruining your clothing. If you spill any acid on you, wash the contaminated area thoroughly and immediately report the incident to your laboratory instructor. You may require further treatment. Be sure to wear goggles.

4. Record the initial buret reading to the nearest 0.01 mL.
5. Place one of the flasks under the buret with the tip inside the mouth of the flask. Place a piece of white paper under the flask.
6. Subtract 2 mL from the volume of the $Na_2S_2O_3$ solution that you calculated in the Prelaboratory Assignment. *Rapidly* add the resulting volume to the flask.
7. Rinse the walls of the flask with distilled water from a plastic wash bottle.
8. Add 40 drops of a 0.2% starch solution.
9. Continue the titration on a *drop-by-drop* basis. Swirl the flask rapidly after each drop. The titration will be finished when one drop causes the solution to become colorless.
10. Record the final buret reading to the nearest 0.01 mL.
11. Refill the buret, if necessary, and repeat Steps 4 through 10. If the volumes that were used in these titrations differ by more than 0.15 mL (about 3 drops), repeat the titration until two consecutive results have this precision.
12. Calculate and record the molarity from each of the two titrations. Obtain the mean molarity.

Analyzing the Saturated Solutions of $Ca(IO_3)_2$

1. The object of this part of the experiment is to determine the molarity of IO_3 ions in the solution above each of the precipitates. Extraneous water must not be introduced during filtration and sampling, or the molarity of IO_3^- ions will no longer be that of a saturated solution.
2. Filter each mixture prepared in the first part of this experiment through a piece of *dry* filter paper and a *dry* filter funnel. Catch each filtrate in a clean, *dry* labeled beaker. Use a different piece of filter paper for each filtration.
3. *Do not wash* the precipitates on the filter paper.
4. Rinse the 10-mL pipet with distilled water, and shake out as much water as you can. Rinse the pipet with two small portions (about 2 mL each) of the first filtrate. Discard these portions.
5. Pipet 10.0 mL of the first filtrate into a clean, labeled 125-mL or 250-mL Erlenmeyer flask.
6. Fill the buret with the $Na_2S_2O_3$ solution.
7. Add about 20 mL of distilled water to each flask from a clean graduated cylinder.
8. Dissolve about 1 cm^3 of solid KI, as measured in a clean, dry 10-mL graduated cylinder, in each solution. Add 20 drops of 2 *M* HCl to each flask, and swirl to obtain homogeneous solutions.

CAUTION: Remember to handle the hydrochloric acid with care. Read the guidelines on handling hydrochloric acid in the section on standardizing the $Na_2S_2O_3$ solution if you have not done so already.

9. Record the initial buret reading to the nearest 0.01 mL.
10. Place one of the flasks under the buret with the tip inside the mouth of the flask. Place a piece of white paper under the flask.
11. Add 40 drops of a 0.2% starch solution.
12. Begin a trial titration by adding increments of about 1 mL of the $Na_2S_2O_3$ solution. Swirl the solution after each addition of the titrant
13. The trial titration is complete when the addition of about 1 mL causes the solution to become colorless. Record the buret reading to the nearest 0.01 mL
14. Repeat Steps 5 through 10 with a second sample of the first filtrate. This is an exact titration.
15. Subtract 1 mL from the volume found in the trial titration. Rapidly add the resulting volume to the flask from the buret.
16. Rinse the walls of the flask with distilled water.
17. Add 40 drops of the starch solution.
18. Continue the titration on a *drop-by-drop* basis, swirling after each drop, until one drop causes the complete disappearance of the color.
19. Record the buret reading to the nearest 0.01 mL
20. Rinse the pipet with two small portions (about 2 mL each) of the second filtrate.
21. Using a 10.0-mL sample from the second filtrate, repeat Steps 5 through 10 and Steps 15 through 19. This is an exact titration.
22. Calculate the molarity of IO_3^- ions in each saturated solution.
23. Dispose of all solutions as directed by your instructor.

Date ______________________ Student Name ______________________
Course/Section ______________________ Team Members ______________________
Instructor ______________________ ______________________

A Solubility Product Constant

Prelaboratory Assignment

1. Provide definitions for the following terms:

   a. Solubility

   b. Saturated solution

   c. Solubility product constant

   d. Standardization

2. a. Give the chemical equation that describes the equilibrium between calcium iodate and its ions in a saturated solution.

   b. Give the chemical equation for the net reaction that occurs during the titration.

3. a. How will you know when a stoichiometric amount of the $Na_2S_2O_3$ solution has been added in the titration?

   b. What is the purpose of the trial titration?

4. You may be required to standardize a solution of $Na_2S_2O_3$ in this experiment. The approximate molarity of this solution will be 0.025 *M*. You will take 10.0 mL of a 0.0100 *M* KIO_3 solution, dissolve excess amounts of KI in it, and, using the $Na_2S_2O_3$ solution, titrate to the disappearance of the starch–I_2 color. Using the approximate molarity of the $Na_2S_2O_3$ solution, calculate the approximate volume of the solution that will be required in the titration.

5. What safety precaution must be observed during this experiment?

Date		Student Name	
Course/Section		Team Members	
Instructor			

A Solubility Product Constant

Results

1. *Standardizing the $Na_2S_2O_3$ solution (optional)*

Sample	1	2	3
KIO_3 solution taken (mL)			
Moles of KIO_3			
Final buret reading (mL)			
Initial buret reading (mL)			
Volume of $Na_2S_2O_3$ solution (mL)			
Molarity of $Na_2S_2O_3$ solution (mol/L)			
Mean molarity (mol/L)			

Calculations:

2. *Analyzing the saturated solutions*

a. Trial titration

Final buret reading (mL): ____________

Initial buret reading (mL): ____________

Volume of $Na_2S_2O_3$ solution (mL): ____________

b. Exact titrations

Filtrate	**1**	**2**
Final buret reading (mL)	____________	____________
Initial buret reading (mL)	____________	____________
Volume of $Na_2S_2O_3$ solution (mL)	____________	____________
Moles of $Na_2S_2O_3$ used	____________	____________
Moles of IO_3^- present initially	____________	____________
Molarity of IO_3^- (mol/L)	____________	____________

Calculations:

Student name: _______________ Course/Section: _______________ Date: _______________

Questions

1. a. Calculate the solubility of $Ca(IO_3)_2$ from your results with the first and second filtrates.

   b. Why should these solubilities be identical? Explain.

   c. Calculate K_{sp} for $Ca(IO_3)_2$, using the mean solubility.

2. How will your calculations be affected by errors introduced from the following sources:

   a. The precipitate of $Ca(IO_3)_2$, when obtained originally, is not washed with distilled water.

   b. The concentration of KIO_3 used for standardizing the solution of $Na_2S_2O_3$ is somewhat greater than 0.0100 *M*.

   c. Extraneous water is introduced from wet filter paper or wet funnels, or by washing the precipitate during the final filtration of $Ca(IO_3)_2$.

17B. Qualitative Analysis of Ag^+, Cu^{2+}, Zn^{2+}, and Ca^{2+} Ions

Introduction

Qualitative analysis involves the identification of the substances in a mixture. Qualitative analysis answers the question "What is in my mixture?" Quantitative analysis answers the question "How much of compound A is in my mixture?" When chemical methods are used in the identification of mixtures of metal cations, these ions are usually separated before identification can occur. After they have been separated, identification of each cation depends on the observation of a characteristic chemical reaction. Solubility equilibria and complex-ion equilibria play crucial roles in the separations and subsequent identifications of these compounds.

Figure 17B.1, on the following page, provides a scheme for qualitative analysis that relies on these equilibria (Ebbing/Gammon, Section 17.7). This experiment provides an introduction to that scheme. Qualitative analysis can also be found in Experiment 21A, "Qualitative Analysis of Mg^{2+}, Ca^{2+}, Ba^{2+}, and Al^{3+} Ions."

Purpose

You will learn to separate and identify each cation in a mixture of Ag^+, Cu^{2+}, Zn^{2+}, and Ca^{2+} ions. You will also receive an unknown mixture for qualitative analysis. This mixture can contain one, two, three or all four of these cations.

Concept of the Experiment

You will be working with mixtures of Ag^+, Cu^{2+}, Zn^{2+}, and Ca^{2+} ions. These cations belong to Analytical Groups I, II, III, and IV, as shown in Figure 17B.1. Time does not permit us to include a cation from Analytical Group V. The intention of this experiment is to give you a better appreciation of this separation scheme through a few examples. Before you try to do the experiment, spend some time studying the scheme and reading the following description.

Hydrochloric acid will be used to precipitate Ag^+ as white AgCl. If you do not observe a precipitate with your unknown mixture, this cation cannot be present. The formation of a precipitate, however, is not considered sufficient evidence for the presence of Ag^+ in either a known or an unknown mixture. To confirm the presence of this cation, you need to verify that this precipitate dissolves in aqueous ammonia with the formation of a complex ion and reappears when the solution is treated with an acid.

FIGURE 17B.1

A diagram for a qualitative-analysis scheme in which metal ions are separated into five analytical groups based on solubility properties of the resulting product.

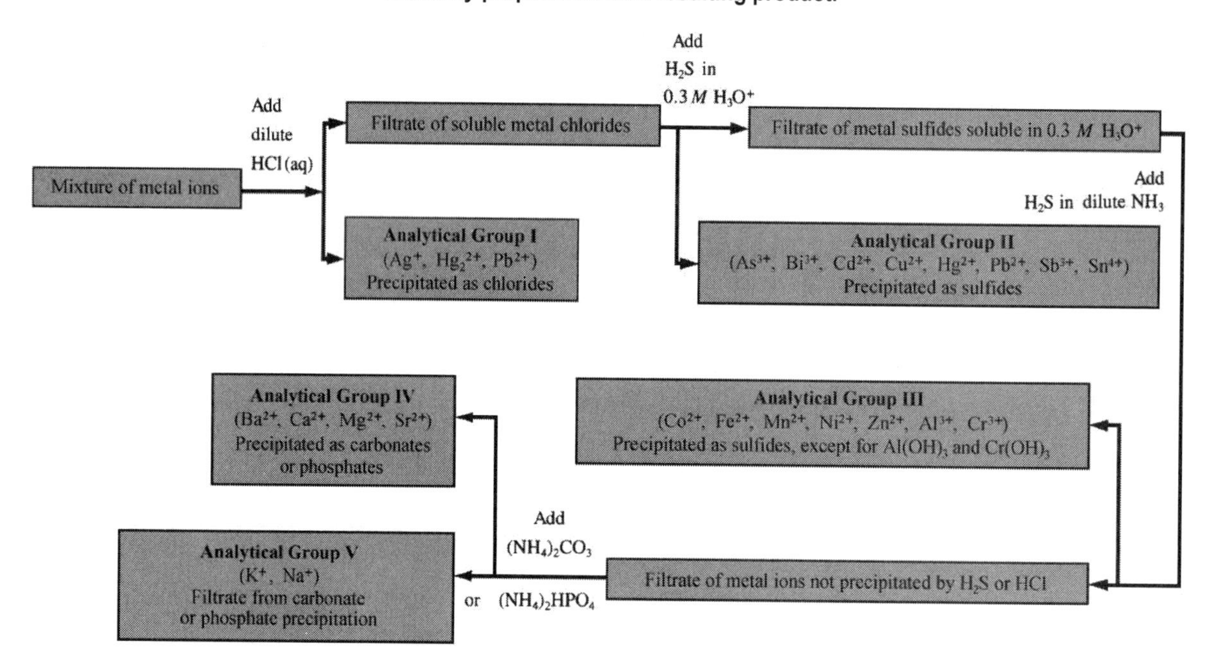

Hydrogen sulfide is required for the next two separations. A saturated solution (0.10 *M*) of this substance will be generated by heating a solution of thioacetamide, CH_3CSNH_2. Hydrolysis (reaction with water) has the following result:

$$CH_3CSNH_2 + H_2O \rightarrow CH_3CONH_2 + H_2S$$

This reagent will be used to precipitate Cu^{2+} as black CuS from a 0.3 $M\ H_3O^+$ solution and to precipitate Zn^{2+} as white ZnS from a weakly basic solution. In the Prelaboratory Assignment, you will find a problem that will show you how to prepare the 0.3 $M\ H_3O^+$ solution. Another problem will show that both CuS and ZnS will precipitate from a weakly basic solution but that only CuS will precipitate from a 0.3 $M\ H_3O^+$ solution. Therefore, separation of these ions will occur only if precipitation is achieved first under acidic conditions and then under basic conditions. If a precipitate does not form in the acidic solution of your unknown mixture, Cu^{2+} must be absent. Similarly, Zn^{2+} cannot be in your unknown mixture if a precipitate is not formed in the weakly basic solution.

You will then dissolve each of these metal sulfides in nitric acid. This reagent will oxidize the sulfide ion to elemental sulfur. The solution should be blue for $Cu^{2+}(aq)$ and colorless for $Zn^{2+}(aq)$. The confirmatory test for each cation involves the addition of potassium ferrocyanide, $K_4Fe(CN)_6$, to these solutions. A red-maroon precipitate confirms the presence of Cu^{2+}, whereas a white precipitate confirms the presence of Zn^{2+}. You should note that $Fe(CN)_6^{4-}$ is a complex ion that is considerably less toxic and dangerous than the cyanide ion, CN^-.

Finally, Ca^{2+} will be precipitated as white $CaCO_3$ by the addition of $(NH_4)_2CO_3$. This precipitate will dissolve in an acid with the evolution of carbon dioxide. If a precipitate does not form, Ca^{2+} cannot be present in your unknown mixture. To confirm the presence ofthis cation you will test for the precipitation of the white oxalate, CaC_2O_4, upon the addition of $K_2C_2O_4$.

Procedure

Getting Started

1. Obtain 6 small test tubes
2. Set up a boiling-water bath using a beaker of water, a ring stand, an iron ring, a wire gauze, and a laboratory burner. Place the bath in a hood if one is available. If not, use an inverted conical filter funnel connected by rubber tubing to a water aspirator. Clamp the funnel so that it will be positioned directly over the test tube in which hydrogen sulfide will be generated (Step 10 of the analysis).

    CAUTION: Avoid burning your fingers. Do not touch the iron ring or wire gauze during heating.

3. Obtain your unknown mixture, and record its identification number and color. Does the color provide a clue about the presence or absence of one of the possible components in the mixture?
4. Obtain 1 mL of the known mixture. This solution contains $AgNO_3$ (0.1 *M*), $Cu(NO_3)_2$ (0.2 *M*), $Zn(NO_3)_2$ (0.2 *M*), and $Ca(NO_3)_2$ (0.2 *M*).
5. Conduct the analysis of the known and unknown solutions simultaneously so that you can compare the results.
6. Use labeled test tubes throughout the experiment so that you do not confuse the known and unknown solutions and precipitates at any time.
7. Obtain instructions for using the centrifuges from your laboratory instructor.

    CAUTION: When you use a centrifuge, do not attempt to stop the centrifuge rotor with your finger or anything else.

8. Obtain instructions for discarding the solutions that you will use in this experiment.
9. Be careful when handling the solutions used in this experiment. Wear gloves if they are available.

    CAUTION: Hydrochloric acid, ammonia, nitric acid, and acetic acid can cause chemical burns, in addition to ruining your clothes. If you spill any of these solutions on you, wash the contaminated area thoroughly with tap water and immediately report the incident to your instructor. You may require further treatment.

Doing the Analysis

1. Take 1 mL of the known mixture and 1 mL of the unknown mixture in separate small test tubes.
2. Each of the subsequent additions and operations should be conducted on both the known and the unknown mixtures unless the instructions indicate not to repeat these steps.
3. Add 2 drops of 6 *M* HCl. If no precipitate forms, proceed with Step 7. If a precipitate forms, stir the mixture with a clean stirring rod. Centrifuge the mixture for about 1 min.
4. Check for complete precipitation by adding 1 more drop of 6 *M* HCl. If you see that additional precipitate has formed, centrifuge the mixture again. Check for complete precipitation by adding 1 more drop of 6 *M* HCl.

    Continue this process until no precipitate is formed. Do not add more than one drop of HCl at a time; you do not want to have an excess of HCl in the solution.

5. Decant (pour off) the solution into a clean test tube. Save this solution for Step 7. Use the precipitate from step 4 in the following step.

6. Add 10 drops of 6 *M* NH_3 to the precipitate. If necessary, stir the mixture with a clean stirring rod until the precipitate dissolves. Add 6 *M* HNO_3 by drops until a white precipitate appears. These two reactions confirm the presence of Ag^+.

7. Add 6 *M* NH_3 by drops to the solution from either Step 3 or Step 5 until a drop of the solution transferred from a clean stirring rod to a piece of pink litmus paper causes the paper to turn blue.

8. Estimate the volume of the solution to the nearest 0.5 mL. Make the estimate by comparing the volume with measured amounts of water poured into a test tube of similar size. Mentally, add 0.6 mL to that volume to account for the dilution that will occur in Step 9. Round the result to the nearest milliliter. Add 1 drop of 6 *M* HCl to the solution for every milliliter that results from this procedure. The concentration of H_3O^+ in the solution will be 0.3 *M* after the dilution in Step 9 is completed.

9. Add 12 drops of 1 *M* thioacetamide to the solution.

   CAUTION: Thioacetamide is a carcinogen. Avoid contact with your skin.

10. Place the test tube in the boiling-water bath for about 10 min. If no precipitate forms during this time, proceed with Step 16. If a black precipitate forms, proceed with the following step.

    CAUTION: In addition to having a foul odor, the hydrogen sulfide generated during the hydrolysis of thioacetamide is extremely toxic. Although only small amounts of H_2S usually escape from the solution, work under a hood if possible. If not, use the inverted conical filter funnel and water aspirator described earlier to suck away the escaping H_2S.

11. Centrifuge the mixture and decant the clear solution into a clean test tube. Save the precipitate for Step 12. Test the solution for complete precipitation by adding 3 more drops of 1 *M* thioacetamide and reheating for 5 min. If no additional precipitate forms, save the solution for Step 16. If a precipitate forms, centrifuge the mixture, decant the clear solution and save it for Step 16, and discard this secondprecipitate in the appropriate waste container.

12. Wash the precipitate from part 1 of step 11 by stirring it vigorously with 1 mL of distilled water. Use a clean stirring rod. Centrifuge the mixture, save the precipitate for step 13, and discard the water.

13. Add 20 drops of 6 *M* HNO_3 to the precipitate from step 12. Place the test tube in the boiling-water bath for several minutes. Stir occasionally with a clean stirring rod. Separate any sulfur or traces of undissolved sulfides by centrifuging the solution. Decant the solution for use in the next step. Discard the solid residue in the appropriate waste container.

14. Add drops of 6 *M* NH_3 carefully until one drop of the solution transferred to a piece of pink litmus paper turns the paper blue. The solution will be basic at this point. Then add drops of 6 *M* acetic acid until one drop of the solution transferred to a piece of blue litmus paper turns the paper pink.

15. Add 10 drops of 0.1 *M* $K_4Fe(CN)_6$ and mix thoroughly. A red-maroon precipitate confirms the presence of Cu^{2+}.

16. Estimate the volume of the solution from either Step 10 or Step 11 by comparing it with 3 mL of water in a test tube of similar size. If the volume is 3 mL or less, proceed to the next step. If it is greater than 3 mL, place the solution in the boiling-water bath until some of the solution has evaporated that volume is attained.

17. Add 10 drops of 6 *M* HCl followed by 10 drops of 6 *M* NH_3. Then add drops of 6 *M* NH_3 until the solution is basic when tested with litmus paper. Add 5 more drops of this reagent.
18. Add 12 drops of 1 *M* thioacetamide. Stir thoroughly and heat for 10 min in the boiling-water bath. If no precipitate forms, proceed with Step 23. If a precipitate forms, proceed with the following step. Follow the same safety precautions as outlined for Steps 9 and 10.
19. Centrifuge the mixture and decant the clear solution into a clean test tube. Save this solution for Step 23. Use the precipitate in the next step.
20. Wash the precipitate by stirring it vigorously with 2 mL of distilled water to which you have added 1 drop of 6 *M* NH_3. Use a clean stirring rod. Centrifuge the mixture, save the precipitate, and discard the water in the appropriate waste container.
21. Repeat Steps 13 and 14.
22. Add 10 drops of 0.1 *M* $K_4Fe(CN)_6$ and mix thoroughly. A white precipitate confirms the presence of Zn^{2+}.
23. Add 10 drops of 3 *M* $(NH_4)_2CO_3$ to the solution from either Step 18 or Step 19. If no precipitate forms, you have completed the analysis. If a white precipitate forms, proceed with the next step.
24. Centrifuge the mixture, save the precipitate, and discard the solution.
25. Wash the precipitate in the same manner as in Step 12.
26. Dissolve the precipitate by adding 5 drops of 6 *M* acetic acid.
27. Make the solution basic to litmus paper using 6 *M* NH_3.
28. Add 10 drops of 1 *M* $K_2C_2O_4$. A white precipitate that should form within a few minutes confirms the presence of Ca^{2+}.

    CAUTION: Wash your hands thoroughly. Oxalate solutions are poisonous.

29. Record the cations that are present in the unknown mixture.

    CAUTION: Before you leave the laboratory, make sure that your gas outlet and those of your neighbors are closed.

ate ______________________ Student Name ______________________
ourse/Section ______________________ Team Members ______________________
structor ______________________ ______________________

Qualitative Analysis of Ag^+, Cu^{2+}, Zn^{2+}, and Ca^{2+} Ions

Prelaboratory Assignment

1. a. What metal cations are involved in this experiment?

   b. How will these cations be separated? What reagents will be used for these separations?

2. Consider a solution of 0.1 *M* Zn^{2+} and 0.1 *M* Cu^{2+} that is saturated with H_2S. You will encounter a similar mixture in this experiment. The solubility product constants for ZnS and CuS are 1.1×10^{-21} and 6×10^{-36}, respectively.

   a. Show that both ZnS and CuS will precipitate at pH 8.

   b. Show that only CuS will precipitate when $[H_3O^+] = 0.3$ *M*.

Student name: _______________ Course/Section: ________________ Date: ______________

3. In this experiment, you will be required to adjust the concentration of H_3O^+ ions to 0.3 *M* so that only the cation from Analytical Group II is precipitated with H_2S. Describe what you will do to attain this concentration of H_3O^+ ions.

4. How will the color of the unknown provide a clue about the presence or absence of one of the possible components of the unknown?

5. What safety precautions must be observed during this experiment?

Date	______________	Student Name	______________
Course/Section	______________	Team Members	______________
Instructor	______________		______________

Qualitative Analysis of Ag^{+}, Cu^{2+}, Zn^{2+}, and Ca^{2+} Ions

Results

Unknown no.: ______________

Color of unknown mixture: ______________________

Ions present: ______________________

Questions

1. Write a balanced equation for each precipitation reaction that was used to separate the four cations.

2. Write a balanced equation for the confirmatory reaction or reactions for each cation.

18. Spontaneity

Introduction

What is the criterion or requirement for the spontaneity of a physical or chemical process? Students often believe that every exothermic process must be spontaneous. However, the sole criterion at constant temperature and pressure (Ebbing/Gammon, Section 18.4) is

$$\Delta H - T\Delta S < 0$$

As you know, ΔH and ΔS are the enthalpy and entropy changes for the process. The entire left-hand side of this inequality is called the *free-energy change*. The symbol for this change is ΔG. There are three rules that are easy to remember. If ΔG is negative ($\Delta G < 0$), the process is spontaneous. If it is positive ($\Delta G > 0$), the process is nonspontaneous. If it is zero, the process is at equilibrium.

Purpose

Using thermochemical methods, you will measure the enthalpy change that occurs when sodium nitrate is dissolved in water. You will also predict the sign of the free-energy change for this process and estimate the minimum value for the entropy change.

Concept of the Experiment

The process that you will examine in this experiment is given by the equation

$$NaNO_3(s) \rightleftharpoons Na^+(aq) + NO_3^-(aq)$$

You will measure the heat evolved or absorbed during this process by using the coffee-cup calorimeter. Appendix B describes this calorimeterand gives definitions of the system and its surroundings in terms of this calorimeter. If you did the experiment "Thermochemistry and Hess's Law" or "A Student's View of Liquids and Solids," you will be familiar with the calorimeter, the technique, and the calculations that give you the enthalpy change for the process. You will pool your data with those of your classmates to obtain better precision in the measured enthalpy change.

After you have obtained ΔH, you will have to decide whether the process is spontaneous or nonspontaneous. The sign of ΔG for the process will rest on your decision. Finally, you should be able to obtain a minimum value for the entropy change (as well as its sign) from ΔH and the predicted sign for ΔG.

Procedure

Getting Started

1. Work with a partner.
2. Obtain a coffee-cup calorimeter.

Measuring the Heat Evolved or Absorbed

1. In the Prelaboratory Assignment, you calculated the mass of $NaNO_3$ that would be required to prepare 100 mL of a 1.0 *M* solution. Using the laboratory balance, weight out this mass of $NaNO_3$. Make sure the pan of the balance is protected by placing the sodium nitrate onto a piece of weighing paper.

2. Place 100 mL of distilled water in the calorimeter, using a clean 100-mL graduated cylinder.
3. Measure and record the temperature of this water to the nearest 0.1°C. This is the initial temperature (t_i).
4. Add the solid $NaNO_3$ to the cup in such a way that none adheres to the side of the cup.
5. Place the top on the calorimeter immediately and begin stirring.
6. Measure the temperature of the solution to the nearest 0.1°C after 30 s and every 30 s thereafter until the temperature attains either a maximum or a minimum value. This temperature will be used as the final temperature (t_f).
7. Calculate q(system), using 4.184 J/(g • °C) and 1.0 g/mL for the specific heat and density of the solution and 1.0×10^1 J/°C for the heat capacity of the calorimeter. (Refer to Appendix B for details on how to do these calculations.)
8. Calculate the enthalpy change, ΔH, from q(system) and the number of moles of $NaNO_3$.
9. Repeat Steps 1 through 8 with a new solution. Calculate the mean enthalpy change for the process.
10. Pool this value with the data obtained by your classmates, and calculate a new mean enthalpy change.

Date	______________	Student Name	______________
Course/Section	______________	Team Members	______________
Instructor	______________		______________

Spontaneity

Prelaboratory Assignment

1. Provide symbols (where appropriate) and definitions for the following terms:

    a. Enthalpy change

    b. Entropy change

    c. Free-energy change

    d. Spontaneous process

    e. Nonspontaneous process

2. a. How will you decide whether a process is spontaneous in this experiment?

b. If you were to use this method after you had observed the fate of an ice cube at 25°C, what would you conclude about the spontaneity of the following process? Why?

$$H_2O(s) \rightleftarrows H_2O(l)$$

\c. The standard enthalpy change for this process is 6.01 kJ/mol. What is the minimum value for the standard entropy change, based on your conclusions about the spontaneity of this process?

3. During this experiment, you will be required to prepare 100 mL of a 1.0 *M* solution of $NaNO_3$. Calculate the mass of $NaNO_3$ (to the nearest tenth of a gram) that will be required.

Date	______________	Student Name	______________
Course/Section	______________	Team Members	______________
Instructor	______________		______________

Spontaneity

Results

Trial	1	2
Mass of $NaNO_3$ and paper (g)	______	______
Mass of paper (g)	______	______
Mass of $NaNO_3$ (g)	______	______
t_i (°C)	______	______
Temperature (°C) after	______	______
30 s	______	______
60 s	______	______
90 s	______	______
120 s	______	______
150 s	______	______
180 s	______	______
210 s	______	______
240 s	______	______
t_f (°C	______	______
(system) (J)	______	______
H (kJ/mol)	______	______
lean ΔH (kJ/mol)	______	

Calculations:

Pooled results (Include your own mean ΔH.)

__________ __________ __________ __________ __________

__________ __________ __________ __________ __________

__________ __________ __________ __________ __________

Mean ΔH: __________

Questions

1. Is the process

$$NaNO_3(s) \rightleftharpoons Na^+(aq) + NO_3^-(aq)$$

spontaneous or nonspontaneous? Why?

Student name: ________________ Course/Section: ________________ Date: ______________

2. a. Use your experimental data and your decision about the spontaneity to calculate a minimum value for the entropy change for the process.

   b. Does the sign of the entropy change predicted by this method agree with the one that you would predict on the basis of the expected change in disorder as solid $NaNO_3$ is dissolved in water? Explain.

3. Many students believe that spontaneous processes must be exothermic. Does your data from this experiment support this theory? Why or why not?

19A. Oxidation–Reduction Reactions

Introduction

We encounter oxidation–reduction reactions every day. Examples include the combustion of coal, natural gas, oil, and gasoline; the operation of an automobile's battery; and the removal of stains by a bleach.

Although there are a variety of oxidation–reduction reactions , when you compare a number of these reactions, you can find some common features. Each reaction consists of simultaneous oxidation and reduction because electrons are lost by one atom and gained by another one. Moreover, the reactants must include an oxidizing agent and a reducing agent. Electrons are always lost by the reducing agent and gained by the oxidizing agent (Ebbing/Gammon, Section 4.5).

Purpose

You will begin by preparing aqueous solutions of three halogens: Cl_2, Br_2, and I_2. You will prepare these substances in solution from stoichiometric quantities of certain oxidizing and reducing agents. By observing the effect of these halogens on the corresponding halides (Cl^-, Br^-, and I^-), you will be able to rank the halogens according to their oxidizing strengths and the halides according to their reducing strengths. Your subsequent observations on the reactions of two halides, Br^- and I^-, with two common oxidizing agents will enable you to rank the latter according to their oxidizing strengths. You will also find that the products obtained from an oxidation–reduction reaction in an acidic solution can differ markedly from those obtained in a basic solution. Finally, and most important, you will write a balanced equation for each reaction that you have observed.

The Halogens and the Halides

You will deal with halogens and halides in each part of this experiment. A brief description of some of their properties will enable you to understand some of your observations more easily.

The halogens are found in Group 7A of the periodic table. The members of this group are fluorine (F), chlorine (Cl), bromine (Br), iodine (I), and astatine (At). Because these elements belong to the same group, they have many similar properties. For example, their elemental form is diatomic (F_2, Cl_2, Br_2, I_2, and At_2), and each of them is an oxidizing agent. This experiment uses only Cl_2, Br_2, and I_2, however, because F_2 is such a strong oxidizing agent that special conditions are required for its study and because astatine is radioactive.

Aqueous solutions of Cl_2, Br_2, and I_2 can be prepared either by dissolving these elements in water or by generating them in solution as products of certain oxidation–reduction reactions. Aqueous solutions of these halogens are called chlorine water, bromine water, and iodine water, respectively. Each solution has its own characteristic color, depending on the concentration and thickness of the solution. Chlorine water is colorless to yellow; bromine water is yellow to red-brown; and iodine water is red-brown to brown.

These halogens are also soluble in cyclohexane (C_6H_{12}), a substance that is itself insoluble in water. When cyclohexane is added to an aqueous solution of a halogen, two layers are formed. The upper layer is cyclohexane, because it is insoluble and less dense than water. A portion of the halogen passes from the aqueous layer to the upper cyclohexane layer (a process called *extraction*). A characteristic color is then imparted to this layer. The color of a halogen in cyclohexane differs somewhat from its color in an aqueous solution, as you will discover during the experiment.

Reduction of a halogen (X_2) yields halide ions (X^-) according to the half-reaction.

$$X_2 + 2e^- \rightarrow 2X^-$$

Because the halogens are oxidizing agents, the halide ions are reducing agents. Compounds of the halides are colorless if the positive ions belong to either Group 1A or Group 2A, as in NaI or $CaCl_2$. Compounds of this type are ionic, so they are insoluble in cyclohexane.

Reduction of Other Common Oxidizing Agents

You will also encounter two other oxidizing agents in this experiment. These are $KMnO_4$ (potassium permanganate) and $FeCl_3$ (iron (III) chloride).

Reduction of purple MnO_4^- can give green MnO_4^{2-} (manganate ion); brown-black MnO_2 (manganese dioxide) which will be a solid; or Mn^{2+}, an ion that is pink but will be virtually colorless at the concentration found in this experiment. Reduction of Fe^{3+} will give Fe^{2+}, a pale green ion that will be virtually colorless at the concentration found in this experiment.

Concept of the Experiment

You will mix aqueous solutions of potential oxidizing and reducing agents in each part of this experiment. However, do not expect a reaction to occur in every mixture that you prepare. If a reaction occurs, there will be certain telltale signals. For example, the formation or disappearance of the characteristic color of one of the halogens in a cyclohexane layer is usually a signal of a reaction in the aqueous layer.

Procedure

Getting Started

1. Obtain 3 large test tubes with stoppers for the halogen solutions, 3 small test tubes, and one piece of fine filter paper.
2. Label the large test tubes so that you will know which halogen each one contains.
3. Obtain directions for discarding the solutions that you use during the experiment from your laboratory instructor.
4. Remember the following safety precaution whenever you use cyclohexane during this experiment:

   CAUTION: Cyclohexane is flammable. No open flames are allowed during this experiment.

Preparing Aqueous Solutions of the Halogens

1. Prepare solutions of Cl_2, Br_2, and I_2 according to the directions that you provided in the Prelaboratory Assignment.
2. Use a balance to obtain the required mass of $Ca(OCl)_2$ on a piece of weighing paper.
3. Prepare each solution in a 125-mL Erlenmeyer flask, using a 10-mL graduated cylinder.

   CAUTION: Halogen gases are toxic. Although only small amounts of halogen gas may escape from the solution, work under a hood if possible. If not, use the inverted conical filter funnel and water aspirator described earlier to suck away any escaping halogen gas.

4. Add the reagents in the following order for chlorine water: (1) reducing agent, (2) acid (0.5 *M* H_2SO_4) if required, (3) water if required, and (4) oxidizing agent.
5. For bromine water, the reaction is slow. The rate depends on the concentrations of the reducing and oxidizing agents and of the acid. Therefore, add these substances and wait 15 min before adding the required amount of water; dilution will retard the rate of reaction.
6. Filter the iodine water using gravity filtration (see the Introduction to this manual) to remove insoluble CuI.
7. Store the halogen solutions in the large test tubes.
8. Place 15 drops of the chlorine water in one of the small test tubes, 15 drops of bromine water in the second, and 15 drops of iodine water in the third. Add about 1 mL of cyclohexane to each test tube, and shake or swirl very gently. No stopper should be required. The color of the halogen should appear in the upper layer. View this layer against a white background. Record each color.
9. Discard the solutions following the directions obtained from your laboratory instructor.

Attempting Reactions of Halogens with Halides

1. Wash and dry the small test tubes.
2. Place 10 drops of 0.2 *M* NaCl in each of 2 test tubes.
3. Add 10 drops of bromine water to the first test tube and 10 drops of iodine water to the second one.
4. Add about 1 mL of cyclohexane to each of these test tubes, shake or swirl gently, and record your observations. If no reaction occurs, only the color of the halogen that you added will appear in the upper layer. Discard the solutions as directed.
5. Wash and dry the test tubes.
6. In an identical fashion, test 10 drops of 0.2 *M* NaBr with chlorine water and iodine water. Record your observations.
7. Finally, after washing and drying the test tubes, test 5 drops of 0.4 *M* NaI with chlorine water and bromine water. Record your observations. Discard the solutions as directed.

Attempting Reactions with Other Common Oxidizing Agents

1. Wash and dry the test tubes.
2. Place 15 drops of 0.2 *M* NaBr into each of the test tubes.
3. Add 1 drop of 0.1 *M* $KMnO_4$ to the first test tube and 5 drops of 0.1 *M* $FeCl_3$ to the second test tube.
4. Add 5 drops of 3 *M* H_2SO_4 to each test tube.
5. Add about 1 mL of cyclohexane to each test tube. Shake or swirl the test tubes very gently, and then record your observations for both layers. Discard the solutions as directed. Wash and dry the test tubes.
6. Repeat Steps 1 through 5, using 8 drops of 0.4 *M* NaI instead of 15 drops of 0.2 *M* NaBr.
7. Wash and dry one of the test tubes.
8. Add 8 drops of 0.4 *M* NaI, 5 drops of 6 *M* NaOH, 1 drop of 0.1 *M* $KMnO_4$, and about 1 mL of cyclohexane to this test tube. Shake or swirl very gently, and record your observations for both layers.
9. Dispose of the halogen solutions according to the directions provided by your laboratory instructor.

Date	______________________	Student Name	______________________
Course/Section	______________________	Team Members	______________________
Instructor	______________________		______________________

Oxidation–Reduction Reactions

Prelaboratory Assignment

1. Define the following terms:

   a. Oxidation

   b. Reduction

   c. Oxidizing agent

   d. Reducing agent

2. Give balanced equations that are in accord with the following reactions. You will use these reactions to prepare solutions of the halogens.

   a. The reaction of OCl^- (hypochlorite ion) and Cl^- in acidic solution gives Cl_2 as the only product containing the halogen.

   b. The reaction of BrO_3^- (bromate ion) and Br^- in acidic solution gives Br_2 as the only product containing the halogen.

c. The reaction of Cu^{2+} with I^- in neutral solution gives I_2 and insoluble CuI. A neutral solution does not contain either an acid or a base.

3. Here and on the following page, describe the preparation of 12 mL of a 0.050 *M* solution of each of the halogens, using the reactions in the preceding question. Available to you are solid $Ca(OCl)_2$; 0.20 *M* solutions of NaCl, NaBr, $NaBrO_3$, and $CuSO_4$; a 0.40 *M* solution of NaI; a 0.50 *M* solution of H_2SO_4; distilled water; and a 10-mL graduated cylinder (where volumes can be read to the nearest 0.1 mL). Calculate the required quantity of each reagent and the volume of water that must be added for dilution.

   a. Chlorine water

Student name: ________________ Course/Section: _________________ Date: _______________

b. Bromine water

c. Iodine water

4. What safety precautions must be observed during this experiment?

Date ____________________ Student Name ____________________
Course/Section ____________________ Team Members ____________________
Instructor ____________________ ____________________

Oxidation–Reduction Reactions

Results

Mass of $Ca(OCl_2)_2$ + paper (g): ____________

Mass of paper (g): ____________

Mass of $Ca(OCl)_2$ (g): ____________

1. *Colors of the halogens in cyclohexane*

Cl_2: ____________

Br_2: ____________

I_2: ____________

2. *Reactions of the halogens with halides*

	Cl^-	Br^-	I^-
Cl_2	X	3.	5.
Br_2	1.	X	6.
I_2	2.	4.	X

3. *Reactions of other common oxidizing agents in acidic solution*

	Br^-	I^-
MnO_4^-	7.	9.
Fe^{3+}	8.	10.

4. *Reaction of MnO_4^- with I^- in basic solution*

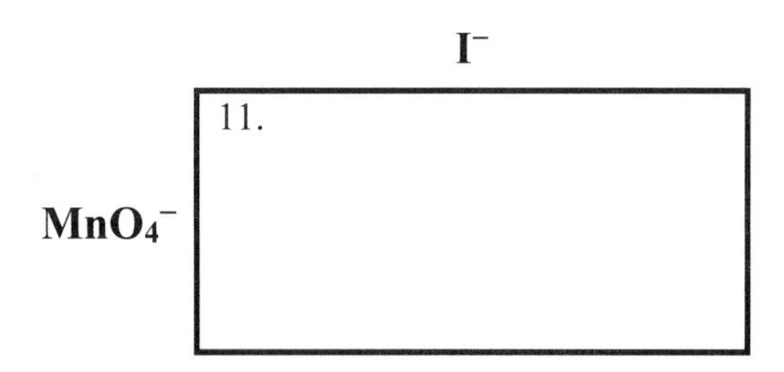

Questions

1. You have made 11 observations. Each one has been numbered in the box in the Results page. This question asks you to consider the first 10 observations. Indicate those cases in which no reaction occurred. If a reaction occurred, write a balanced equation that is in accord with or describes the chemistry resulting in your observation.

(1)

(2)

(3)

Student name: _______________ Course/Section: ________________ Date: ______________

(4)

(5)

(6)

(7)

(8)

(9)

(10)

2. Consider the eleventh observation.

   a. The ultimate product that results from the oxidation of I^- in this reaction is IO_3^-. Although your observations will not prove that this colorless ion was formed, describe how your results indicate that I_2 was *not* formed.

   b. The color of the aqueous layer will enable you to determine without doubt the product from the reduction of MnO_4^-. What is that product?

   c. Write a balanced equation for the reaction.

   d. How do the reactions in acidic and basic solutions differ?

3. a. Rank the oxidizing strengths of the halogens according to your results. Explain how you selected this order.

Student name: ________________ Course/Section: ________________ Date: ________________

b. Rank the reducing strengths of the halides. Explain how you determined this order.

c. Rank the oxidizing strengths of MnO_4^- and Fe^{3+}. Explain your ranking.

19B. Electrochemistry

Introduction

A study of *electrochemistry* (Ebbing/Gammon, Chapter 19) generally consists of at least two principal topics. The first of these deals with the electricity obtained from spontaneous chemical reactions occurring in *voltaic cells.* Batteries are examples of voltaic cells used every day. The second concerns the application of electricity to drive chemical reactions within *electrolytic cells* in nonspontaneous directions. Electrolytic cells have been used in chemical manufacturing to produce hydrogen gas.

Purpose

In this experiment you will study an electrolytic cell and several voltaic cells, including a concentration cell.

The Design of Electrochemical Cells

A voltaic cell functions only when a complete electric circuit occurs. An external wire provides a path for electrons to travel from the anode to the cathode. The circuit is completed when the solutions surrounding these electrodes are in contact. Ions can then pass from one solution to the other. Because several different ways of achieving this contact have been devised, there are several designs for voltaic cells. (Figure 19B.1)

A salt bridge, a porous pot, a porous glass plug, or a membrane can be used to separate the solutions while maintaining electrical contact. These designs are shown in Figure 19B.1. The voltaic cell that you encounter in this experiment will probably be one of these designs.

Electrolytic cells may differ from voltaic cells in one important respect. The solutions around the two electrodes do not always need to be separated. One solution may be enough. An example involving the electrolysis of water is shown in Figure 19B.2. In this chemical reaction, electrical energy added to water breaks the OH bond in water and produces hydrogen and oxygen gas.

Using a Voltmeter

The voltmeter that you use in this experiment may be similar to one shown in Figure 19B.1. Alternatively, it could be a pH meter operating as a voltmeter. Either of these instruments will enable you to measure the voltage from a voltaic cell. Moreover, either will allow you to identify the voltaic cell's anode and cathode.

You will find two terminals on your voltmeter. One of these (you will determine which one) must always be connected to the anode of any voltaic cell, and the other must always be connected to the cathode. The deflection of the voltmeter's needle will be positive only when the correct terminal is connected to the appropriate electrode. If the connections are reversed, a negative deflection will occur. You will use a zinc–copper voltaic cell (Figure 19B.1; also Ebbing/Gammon, Section 19.2) to decide how the terminals must be connected. This information will enable you to identify the anode and cathode of any other voltaic cell.

FIGURE 19B.1

A zinc–copper voltaic cell with (A) a salt bridge, (B) a porous pot, (C) a porous glass plug, and (D) a membrane. Both ions and water can pass through the membrane.

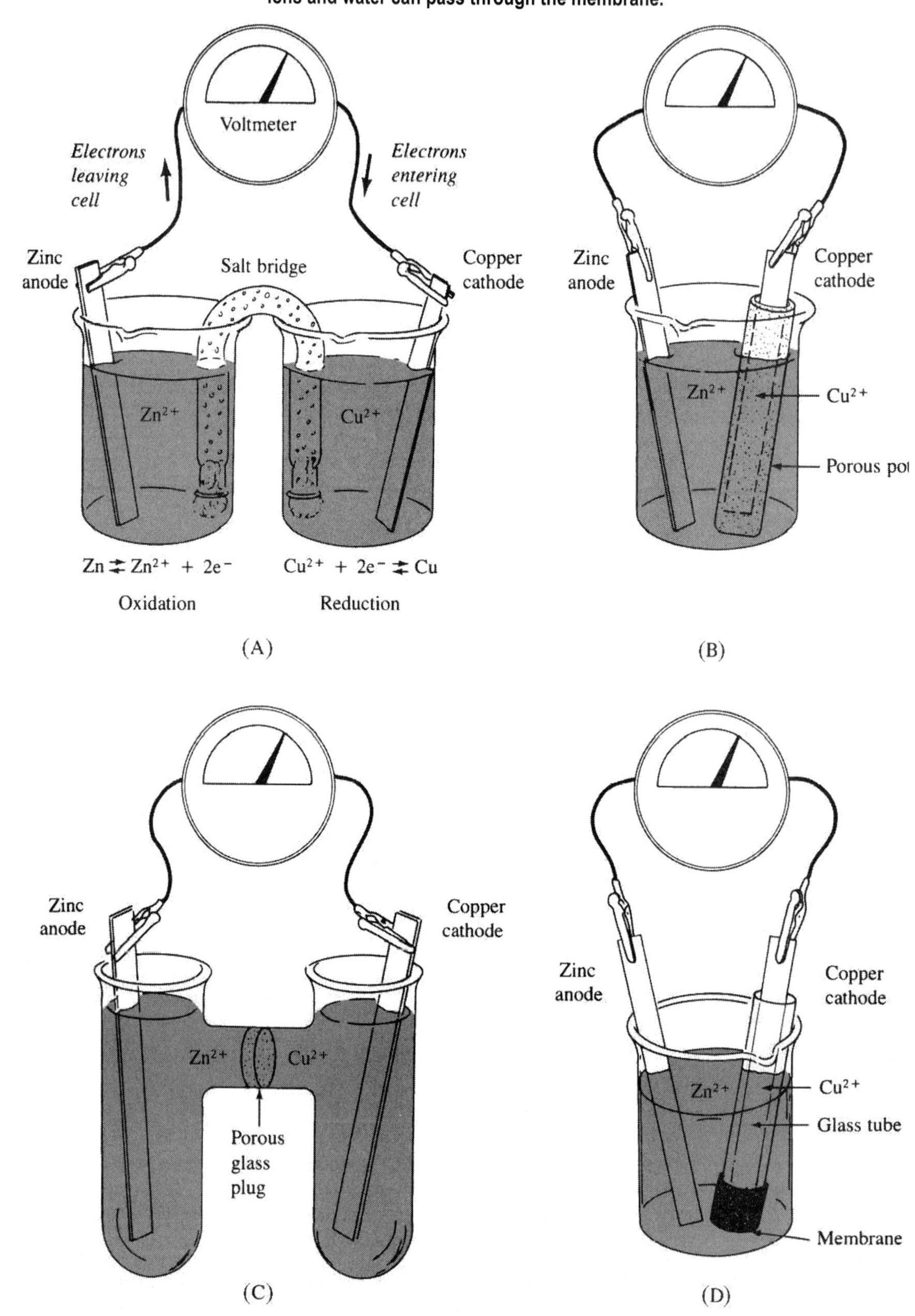

FIGURE 19B.2

The electrolysis of water. Note that the electrodes are not separated by a porous barrier.

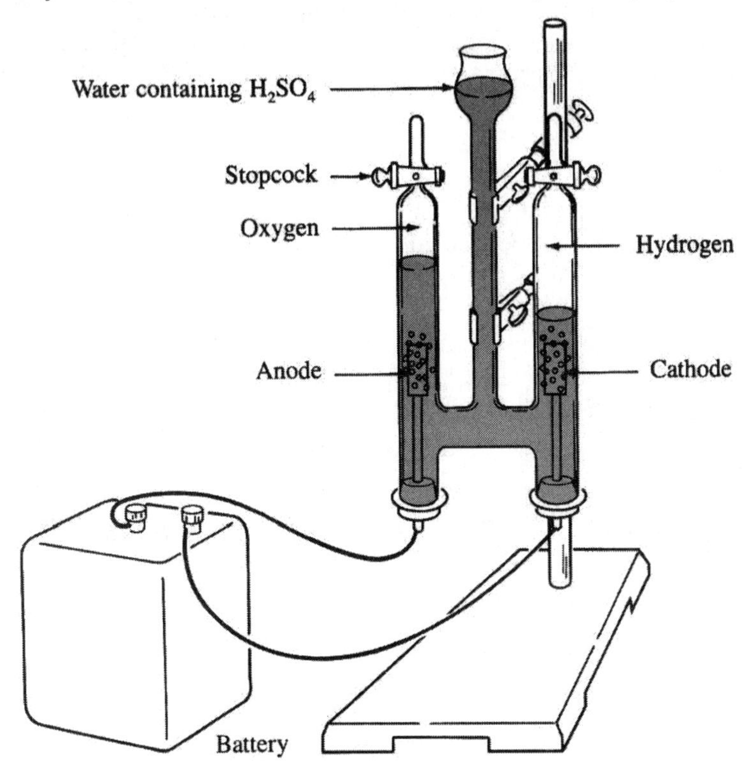

Concentration Cells

A concentration cell is a voltaic cell in which the two half-reactions are identical but involve solutions at different concentrations. Take a lead concentration cell, for example. The half-reactions for this cell are

$$Pb(s) \rightarrow Pb^{2+}(C_1) + 2e^- \qquad E° = +0.13 \text{ V}$$

$$Pb^{2+}(C_2) + 2e^- \rightarrow Pb(s) \qquad E° = -0.13 \text{ V}$$

where C_1 and C_2 are the concentrations of Pb^{2+} ions around each electrode and E° is the standard reduction potential of the half reaction. The overall cell reaction is

$$Pb^{2+}(C_2) \rightarrow Pb^{2+}(C_1)$$

If $C_1 = C_2$, the cell potential will be exactly zero no matter what values C_1 and C_2 actually have. Cell potential will develop, however, when the concentrations of Pb^{2+} ions around the electrodes are different ($C_1 \neq C_2$). We can also say that the overall cell reaction will be spontaneous in the direction inwhich it is written when $C_2 > C_1$.

What makes this concentration cell work? It works because the process

$$Pb^{2+}(C_2) \rightarrow Pb^{2+}(C_1)$$

is spontaneous whether it occurs inside or outside of a voltaic cell. The following example should help you understand the spontaneity of this process when it occurs outside of a voltaic cell. Suppose we have two solutions of Pb^{2+} ions separated by a porous barrier. The concentrations of Pb^{2+} ions in these solutions will again be represented by C_1 and C_2. If $C_2 > C_1$, the natural tendency is for ions to flow through the barrier in such a way that C_2 decreases and C_1 increases. Ions will continue to flow from one side of the barrier to the other until the concentrations of the solutions become equal. This process, which will occur without any outside intervention, is clearly spontaneous. As a result, it must also be spontaneous in the voltaic cell.

What potential will be delivered by a concentration cell? The potential can be calculated from the Nernst equation (Ebbing/Gammon, Section 19.7). At 25°C, this equation is

$$E_{\text{cell}} = E^{\circ}_{\text{cell}} - \frac{0.0592}{n}\log Q$$

where n is the number of electrons transferred in the overall cell reaction and Q is the reaction quotient (Ebbing/Gammon, Section 19.7). For the lead concentration cell, we have

$$E^{\circ}_{\text{cell}} = 0.13 - 0.13 = 0.00 \text{ V}$$

$$Q = C_1/C_2$$

If $C_1 = 0.010$ M and $C_2 = 1.0$ M, the cell potential will be

$$E_{\text{cell}} = -\frac{0.0592}{2}\log\frac{0.010}{1.0} = 0.059 \text{ V}$$

because $n = 2$ for the lead concentration cell. If, however, $C_1 = 0.00010$ M and $C_2 = 1.0$ M again, the cell potential will increase, so that its value is

$$E_{\text{cell}} = -\frac{0.0592}{2}\log\frac{0.00010}{1.0} = 0.118 \text{ V}$$

These examples show that the cell potential of a concentration cell increases as the disparity between the concentrations of the two solutions increases.

Concept of the Experiment

The experiment has four parts. When you have completed the entire experiment, you will have gained valuable insights into some of the fundamental aspects of electrochemistry.

In the first part, you will study electrolysis. A source of direct current, such as a 6-V battery, will provide the electricity. The electrolytic cell will consist of two graphite electrodes immersed in a solution of potassium iodide. The overall cell reaction will be

$$2I^-(aq) + 2H_2O(l) - I_2(aq) + 2OH^-(aq) + H_2(g)$$

You will measure the pH of the solution before the electrolysis. After the electrolysis begins, you will be able to see the evolution of gas at one electrode and the appearance of a yellow-brown color at the other electrode. This color is due to aqueous iodinc. Aftcr 15 min have elapsed, you will make a final pH measurement, and from this you can determine the extent of this reaction and the average current that passed through the electrolytic cell.

In the second part, you will assemble a zinc–copper voltaic cell. You will determine the proper way to connect the voltmeter to the voltaic cell while you are measuring the cell's potential.

In the third part, you will determine the anodes, cathodes, and voltages of a series of voltaic cells. Each cell will have two of the following electrodes: Cu^{2+} | Cu, Fe^{2+} | Fe, Pb^{2+} | Pb, and Zn^{2+} | Zn, as well as Fe^{3+}, Fe^{2+} | Pt (graphite may be used in place of platinum if necessary). You will be able to compare the potential from each cell with a cell potential that you will calculate from the data in Table 19B.1.

Finally, you will examine a copper concentration cell. You will measure the cell potential and study the effect of altering the concentration of Cu^{2+} ions around one electrode.

Procedure

Getting Started

1. Your laboratory instructor may ask you to work with one or two other students.
2. Obtain enough glassware and equipment to assemble a voltaic cell and an electrolytic cell.
3. Obtain a source of direct current.
4. If a salt bridge is required, ask your laboratory instructor about its composition.
5. Obtain the zinc, copper (2), lead, iron, graphite (2), and platinum (or graphite) electrodes and a voltmeter.
6. If your voltmeter is a pH meter, your laboratory instructor may want to give you some instructions in its use.
7. Locate a pH meter that can be used to measure pH. This instrument should be standardized at pH 10.0.
8. Obtain directions for discarding the solutions that you will use in this experiment from your laboratory instructor.

Doing the Electrolysis

1. Obtain 2.0 g of solid KI, using a laboratory balance. Do not place this solid directly on the pan of the balance as it will corrode the metal. Use a weighing paper to protect the balance.
2. Transfer the solid to a clean 125-mL or 250-mL Erlenmeyer flask. Add 100 mL of distilled water from a graduated cylinder, and swirl until a homogeneous solution is obtained.
3. Measure and record the pH of this solution. When you have an opportunity, calculate and record the hydroxide ion concentration, $[OH^-]$.
4. Place two graphite electrodes in a 150-mL beaker. Clamp the electrodes so that they cannot touch each other. Your laboratory instructor will tell you how to do this if it is not obvious from your equipment.

Table 19B.1

Standard Electrode (Reduction) Potentials in Aqueous Solution at 25°C*

Cathode (Reduction) Half-Reaction	Standard Potential, $E°$ (V)
$Li^+(aq) + e^- \rightleftharpoons Li(s)$	–3.04
$K^+(aq) + e^- \rightleftharpoons K(s)$	–2.92
$Ca^{2+}(aq) + 2e^- \rightleftharpoons Ca(s)$	–2.76
$Na^+(aq) + e^- \rightleftharpoons Na(s)$	–2.71
$Mg^{2+}(aq) + 2e^- \rightleftharpoons Mg(s)$	–2.38
$Al^{3+}(aq) + 3e^- \rightleftharpoons Al(s)$	–1.66
$2H_2O(l) + 2e^- \rightleftharpoons H_2(g) + 2OH^-(aq)$	–0.83
$Zn^{2+}(aq) + 2e^- \rightleftharpoons Zn(s)$	–0.76
$Cr^{3+}(aq) + 3e^- \rightleftharpoons Cr(s)$	–0.74

continued

Cathode (Reduction) Half-Reaction	Standard Potential, $E°$ (V)
$Fe^{2+}(aq) + 2e^- \rightleftharpoons Fe(s)$	−0.41
$Cd^{2+}(aq) + 2e^- \rightleftharpoons Cd(s)$	−0.40
$Ni^{2+}(aq) + 2e^- \rightleftharpoons Ni(s)$	−0.23
$Sn^{2+}(aq) + 2e^- \rightleftharpoons Sn(s)$	−0.14
$Pb^{2+}(aq) + 2e^- \rightleftharpoons Pb(s)$	−0.13
$Fe^{3+}(aq) + 3e^- \rightleftharpoons Fe(s)$	−0.04
$2H^+(aq) + 2e^- \rightleftharpoons H_2(s)$	0.00
$Sn^{4+}(aq) + 2e^- \rightleftharpoons Sn^{2+}(aq)$	0.15
$Cu^{2+}(aq) + e^- \rightleftharpoons Cu^+$	0.16
$Cu^{2+}(aq) + 2e^- \rightleftharpoons Cu(s)$	0.34
$IO^-(aq) + H_2O + 2e^- \rightleftharpoons I^-(aq) + 2OH^-(aq)$	0.49
$I_2(s) + 2e^- \rightleftharpoons 2I^-$	0.54
$Fe^{3+}(aq) + e^- \rightleftharpoons Fe^{3+}(aq)$	0.77
$Hg_2^{2+}(aq) + 2e^- \rightleftharpoons 2Hg(l)$	0.80
$Ag^+(aq) + e^- \rightleftharpoons Ag(s)$	0.80
$Hg^{2+}(aq) + 2e^- \rightleftharpoons Hg(l)$	0.85
$ClO^-(aq) + H_2O(l) + 2e^- \rightleftharpoons Cl^-(aq) + 2OH^-(aq)$	0.90
$2Hg^{2+}(aq) + 2e^- \rightleftharpoons Hg_2^{2+}(aq)$	0.90
$NO_3^-(aq) + 4H^+(aq) + 3e^- \rightleftharpoons NO(g) + 2H_2O(l)$	0.96
$Br_2(l) + 2e^- \rightleftharpoons 2Br^-(aq)$	1.07
$O_2(g) + 4H^+(aq) + 4e^- \rightleftharpoons 2H_2O(l)$	1.23
$Cr_2O_7^{2-}(aq) + 14H^+(aq) + 6e^- \rightleftharpoons 2Cr^{3+}(aq) + 7H_2O(l)$	1.33
$Cl_2(g) + 2e^- \rightleftharpoons 2Cl^-(aq)$	1.36
$MnO_4^-(aq) + 8H^+(aq) + 5e^- \rightleftharpoons$ Mn2+$(aq) + 4H_2O(l)$	1.49
$H_2O_2(aq) + 2H^+(aq) + 2e^- \rightleftharpoons 2H_2O(l)$	1.78
$S_2O_8^{2-}(aq) + 2e^- \rightleftharpoons 2SO_4^{2-}(aq)$	2.01
$F_2(g) + 2e^- \rightleftharpoons 2F^-(aq)$	2.87

* See Ebbing/Gammon Appendix I for a more extensive table.

5. Add 30 mL of the solution of KI to the beaker from a graduated cylinder.
6. Attach wires from the battery (or other source) to the electrodes. Immediately note and record the time.
7. Observe the solution for several minutes. What do you see? Record your observations.
8. Allow the electrolysis to proceed for about 15 min. During this time, you can begin your studies of voltaic cells.

9. After about 15 min has elapsed, disconnect the wires, noting the time to the nearest minute.

   Record the time.

10. Remove the electrodes from the solution and examine them thoroughly. What do you see? Record your observations and the probable identity of any substance that may be present.

11. Stir the solution thoroughly, and then measure the pH. When you have an opportunity, calculate and record the hydroxide concentration, $[OH^-]$.

12. Calculate and record the change in $[OH^-]$ that occurred during the electrolysis.

Studying the Zinc–Copper Voltaic Cell

1. Place 0.10 *M* $Cu(NO_3)_2$ in one compartment of a clean voltaic cell and 0.10 *M* $Zn(NO_3)_2$ in the other compartment.

2. Clean the copper and zinc electrodes by dipping them briefly into about 40 mL of 6 *M* HNO_3. Their surfaces will become bright and lustrous. Wash the electrodes with distilled water and dry them.

   CAUTION: Handle the 6 *M* solution of nitric acid carefully. The acid can cause chemical burns, in addition to ruining your clothes. If you get any of it on you, wash the contaminated area thoroughly and immediately report the incident to your laboratory instructor. You may require further treatment.

3. Insert the copper electrode in the solution of $Cu(NO_3)_2$, and insert the zinc electrode in the solution of $Zn(NO_3)_2$.

4. Connect each electrode to a terminal on the voltmeter.

5. If a negative deflection of the voltmeter's needle occurs, the connections should be reversed.

6. Use a piece of marked tape to identify the terminal on the voltmeter that must be connected to the anode. Use another piece of marked tape to identify the terminal that must be connected to the cathode. You identified the anode and cathode of this cell in the Prelaboratory Assignment.

7. Record the cell potential.

8. Disconnect the wires from the electrodes of the voltaic cell.

Studying the Other Voltaic Cells

1. You may begin this part of the experiment while the electrolysis is progressing. These studies can be interrupted when the electrolysis is completed.

2. After you have read all of the instructions, determine the anodes, cathodes, and voltages of the following voltaic cells:

   a. Cu^{2+} | Cu and Pb^{2+} | Pb

   b. Cu^{2+} | Cu and Fe^{2+} | Fe

   c. Cu^{2+} | Cu and Fe^{3+}, Fe^{2+} | Pt (or graphite)

   d. Zn^{2+} | Zn and Pb^{2+}|Pb

   e. Zn^{2+}|Zn and Fe^{2+} | Fe

   f. Zn^{2+} | Zn and Fe^{3+}, Fe^{2+} | Pt (or graphite)

   g. Fe^{2+} | Fe and Fe^{3+}, Fe^{2+} | Pt (or graphite)

If you are using a salt bridge, also include

h. Pb^{2+} | Pb and Fe^{2+} | Fe

i. Pb^{2+} | Pb and Fe^{3+}, Fe^{2+} | Pt (or graphite)

You should not study these two cells if you are using a porous pot, a porous glass plug, or a membrane because the precipitation of $PbSO_4$, which can occur at the junction of the solutions, would plug the pores of these pieces of equipment and cause errors in your results.

3. The solutions that are to be used for these voltaic cells are 0.10 *M* solutions of $Zn(NO_3)_2$, $Pb(NO_3)_2$, $FeSO_4$ (stored over iron wire or nails), and $Cu(NO_3)_2$, along with an iron solution composed of 0.050 *M* $FeSO_4$ and 0.050 *M* $Fe_2(SO_4)_3$ in 1 *M* H_2SO_4.

4. Prepare the copper, iron, lead, and zinc electrodes by cleaning, rinsing, and drying them according to the directions given previously.

CAUTION: When you have completed this part of the experiment, wash your hands. Solutions containing lead are poisonous.

Studying a Copper Concentration Cell

1. Add 1 mL of 0.10 *M* $Cu(NO_3)_2$ from a 5-mL or 10-mL graduated cylinder to a clean, dry 100-mL graduated cylinder. Add enough distilled water to bring the volume up to the 100-mL mark. Pour this solution into a clean beaker, and swirl or stir to obtain a homogeneous solution.
2. Clean the surfaces of two copper electrodes with 6 *M* HNO_3. Wash the electrodes with distilled water and dry them.
3. Place an appropriate quantity of the diluted solution from Step 1 in one compartment of a voltaic cell and an appropriate quantity of 0.10 *M* $Cu(NO_3)_2$ in the other compartment.
4. Insert the copper electrodes into these solutions.
5. Use the voltmeter to identify the anode and cathode and to measure the cell potential. Record your results.
6. Add 10 drops of 0.10 *M* $Cu(NO_3)_2$ to the more dilute solution in the cell. Stir and repeat Step 5.
7. Remove the tape from the voltmeter when you are sure that you do not need to repeat any part of this experiment.

Date	____________	Student Name	____________
Course/Section	____________	Team Members	____________
Instructor	____________		____________

Electrochemistry

Prelaboratory Assignment

1. Provide definitions for the following terms:

   a. Anode

   b. Cathode

   c. Voltaic cell

   d. Electrolytic cell

   e. Concentration cell

2. a. What is the overall cell reaction of the zinc–copper voltaic cell?

   b. Identify which elements are at the anode and cathode.

3. a. What is the overall cell reaction in the electrolytic cell used in this experiment?

   b. Identify the elements at the anode and cathode. Give the half-reactions that will occur at these electrodes.

4. Identify the half-cell reactions for the electrodes that you will study in the third part of this experiment.

5. a. What is the overall cell reaction of a copper concentration cell?

Student name: _______________ Course/Section: ________________ Date: ______________

b. Why is this a spontaneous reaction?

c. Identify the anode and the cathode. Be specific in giving your answer because both electrodes are identical metals, only the concentration of the solution varies

6. What safety precautions are cited in this experiment?

Date ______________________ Student Name ______________________
Course/Section ______________________ Team Members ______________________
Instructor ______________________ ______________________

Electrochemistry

Results

1. *Electrolysis*

Initial pH: ____________

Initial $[OH^-]$: ____________ *M*

Time begun: ____________

Observations:

Time ended: ________________

Time elapsed: ________________

Condition of electrodes after the electrolysis:

Final pH: ________________

Final $[OH^-]$: ________________ *M*

$\Delta[OH^-]$: ________________ *M*

2. *Zinc–copper voltaic cell*

cell potential: ________ V

3. *Other voltaic cells*

Electrodes		Cell Potential (V)	Anode	Cathode

4. *Copper concentration cell*

Anode: _____________

Cathode: _____________

cell potential: _________V

cell potential after 10 drops of 0.10 M $Cu(NO_3)_2$: _________V

Student name: ______________ Course/Section: ______________ Date: ______________

Questions

1. Calculate the average current that flowed through the electrolytic cell from $\Delta[OH^-]$, the volume of the solution, and the time of electrolysis.

2. a. Use the identities of the anode and cathode to give the overall cell reaction for each voltaic cell that you studied after the zinc–copper voltaic cell.

b. Use the Nernst equation, where necessary, to calculate E°_{cell} for each overall cell reaction. Include the zinc–copper voltaic cell.

3. Complete the following table, using your results from every voltaic cell, including the zinc–copper voltaic cell.

Overall Reaction in Voltaic Cell	Calculated E°_{cell} (V)*	Expected E°_{cell} (V)

* From calculations using data in Table 19B.1.

Student name: ________________ Course/Section: ________________ Date: ________________

4. Show that your experimental results are in agreement with the following statements about reactions that occur outside of a voltaic cell.

   a. Zn will reduce Cu^{2+} ions, but Zn^{2+} ions are not reduced by Cu.

   b. Fe^{2+} ions are not oxidized by Zn^{2+} ions.

   c. When an iron nail is inserted into a solution of Cu^{2+} ions, metallic copper is deposited on the nail.

   d. When a piece of lead is inserted into a solution of Zn^{2+} ions, no reaction occurs.

5. a. Calculate the expected cell potential from the copper concentration cell before the addition of 0.10 *M* $Cu(NO_3)_2$.

b. Compare the calculated cell potential with the observed cell potential. Provide a reasonable explanation of any major disagreement.

c. How did the addition of 0.10 *M* $Cu(NO_3)_2$ affect the cell potential? Why?

19C. Estimating an Electrode Potential

Introduction

The reaction in the zinc–copper voltaic cell (Ebbing/Gammon, Section 19.2) occurs spontaneously because $\Delta G < 0$. Metallic zinc is oxidized at one electrode with the formation of Zn^{2+} ions. Simultaneously, Cu^{2+} ions are reduced at the other electrode, and metallic copper is deposited. Spontaneous oxidation–reduction reactions such as this one are not confined to voltaic cells. When a piece of zinc is placed in a beaker containing a solution of Cu^{2+} ions, an identical reaction occurs. Both reactions occur because they have favorable and, of course, identical changes in free energy.

We can see from this example that an oxidation–reduction reaction that is spontaneous in a single container, such as a beaker or flask, will also be spontaneous in a voltaic cell. Moreover, we can reasonably conclude that a reaction that is not spontaneous in a single container will not be spontaneous in a voltaic cell. Behavior in a voltaic cell must be identical to behavior in a single container.

Often you can learn some of the characteristics of a particular electrode by observing its reactions with other substances in a single container. It is even possible to estimate the magnitude of the electrode potential of this electrode without using a voltmeter.

Purpose

You will estimate the standard electrode potential for the $VO_2^+(aq)$, $VO^{2+}(aq)$, $H^+(aq)$ | Pt electrode in this experiment. The results from the interaction of potential reducing agents with VO_2^+ ions and potential oxidizing agents with VO^{2+} ions will enable you to accomplish this goal.

Estimating E°

The standard electrode potential (Ebbing/Gammon, Section 19.5) for the reduction of a substance can sometimes be estimated without recourse to a voltaic cell or a voltmeter. The estimate is based on observation of the spontaneity of the substance's reactions with a series of other substances whose standard electrode potentials are known.

Suppose, for example, that we want to use this technique to estimate the standard electrode potential for the electrode

$$Cu^{2+}(aq) \mid Cu(s) \qquad E° = x$$

How should we proceed?

First, we might choose to explore the reaction

$$Cu(s) + Br_2(l) \rightarrow Cu^{2+}(aq) + 2Br^-(aq)$$

The appropriate half-cell reactions for this overall reaction are

$$Cu(s) \rightarrow Cu^{2+}(aq) + 2e^-$$

$$Br_2(l) + 2e^- \rightarrow 2Br^-(aq)$$

If this reaction were conducted in an electrochemical cell, the electrodes and standard electrode potentials that correspond to these half-cell reactions would be

$$Cu^{2+}(aq) \mid Cu(s) \qquad E° = x$$

$$Br^-(aq) \mid Br_2(aq) \mid Pt \qquad E = 1.07\text{ V}$$

The anode can be established readily because it would be the site of oxidation. The overall reaction shows that metallic copper is oxidized. The cathode must then be the bromide electrode. The standard cell potential is

$$E^\circ_{cell} = E^\circ_{cathode} - E^\circ_{anode}$$
$$= (1.07 - x)\text{V}$$

When we try to do this reaction, we find that it occurs readily. Consequently, it is spontaneous, and E°_{cell} must be positive (that is, $E^\circ_{cell} > 0$). We can then write

$$E^\circ_{cell} = (1.07 - x)\text{V} > 0$$

If $1.07 - x$ is greater than zero, x must be less than 1.07. As a result, we can write

$$x < 1.07\ \text{V}$$

Next, we might choose to explore the reaction

$$Cu(s) + 2H^+(aq) \rightarrow Cu^{2+}(aq) + H_2(g)$$

If this reaction were conducted in a voltaic cell, the electrodes and standard electrode potentials would be

$Cu^{2+}(aq) \mid Cu(s)$	$E^\circ = x$
$HO^+(aq) \mid H(g) \mid Pt$	$E^\circ = 0.00\ \text{V}$

The location of the anode can again be obtained from the overall reaction. Metallic copper is oxidized, so the copper electrode is once again the anode. Now we can write

$$E^\circ_{cell} = E^\circ_{cathode} - E^\circ_{anode}$$
$$= (0 - x)\text{V} = -x\ \text{V}$$

Experiments, however, would show us that metallic copper will not dissolve in strong acids, such as HCl or H_2SO_4. Clearly, the overall reaction is not spontaneous, and E°_{cell} must be negative ($E^\circ_{cell} < 0$). We can then write

$$E^\circ_{cell} = -x\ \text{V} < 0$$

Because $-x$ is less than zero, x must be greater than zero, or

$$x > 0\ \text{V}$$

Thus two experiments have been sufficient to show that x must lie between 0 and 1.07 V, or $0 < x < 1.07$ V. However, this range is very broad. Additional observations with other reagents will enable us to narrow the permissible range.

Suppose that we next elect to examine the reaction

$$Cu^{2+}(aq) + 2Ag(s) \rightarrow Cu(s) + 2Ag^+(aq)$$

The electrodes and their standard potentials are

$Cu^{2+}(aq) \mid Cu(s)$	$E^\circ = x$
$Ag^+(aq) \mid Ag(s)$	$E^\circ = 0.80\ \text{V}$

If this reaction occurs, metallic silver will be oxidized and Cu^{2+} ions will be reduced. We then know that the silver electrode is the anode and the copper electrode is the cathode. The standard cell potential of the voltaic cell is

$$E^\circ_{cell} = E^\circ_{cathode} - E^\circ_{anode}$$
$$= (x - 0.80)\ V$$

An experiment would show that this reaction will not occur. We then know that

$$E^\circ_{cell} = (x - 0.80)\ V < 0$$

If $x - 0.80$ is less than zero, x must be less than 0.80, or

$$x < 0.80\ V$$

This observation enables us to decrease the permissible range for x from $0 < x < 1.07$ V to $0 < x < 0.80$ V. Additional studies with other reagents would lead to a range that spanned only one or two tenths of a volt. The correct value for x is 0.34 V.

A Special Note

A standard electrode potential is the potential of an electrode with the molarities of ions and the pressure of gases (in atmospheres) equal to 1 at 25°C (Ebbing/Gammon, Section 19.5). These conditions do not have to be duplicated faithfully, however, if the standard electrode potential is to be estimated rather than measured directly. The reason is straightforward. As we have seen, the estimated potential will span at least one or two tenths of a volt. Even this limited range will introduce an uncertainty that will exceed the usual deviation from the standard electrode potential caused by nonstandard conditions. As a result, exact conformity to standard conditions and corrections using the Nernst equation (Ebbing/Gammon, Section 19.7) are usually pointless.

Reaction Rate vs. Spontaneity

Spontaneous reactions may or may not occur rapidly. Some spontaneous reactions may be so slow that for all practical purposes, they do not occur under ordinary conditions. Unfortunately, there is no easy way to decide whether a spontaneous reaction will be fast or slow.

Unless you are cautious, a very slow reaction can lead to an incorrect result when you are estimating an electrode potential. For example, when metallic copper is placed in an acidic solution that is saturated with O_2, there is no immediate reaction. If the reaction occurs eventually, we write

$$2Cu(s) + O_2(g) + 4H_3O^+(aq) \rightarrow 2Cu^{2+}(aq) + 6H_2O(l)$$

In a voltaic cell, the electrodes and their standard potentials would be

$$Cu^{2+}(aq)\ |\ Cu(s) \qquad E^\circ = x$$
$$HO^+(aq)\ |\ O(g)\ |\ Pt \qquad E^\circ = 1.23\ V$$

Once again, the copper electrode is the anode, because oxidation occurs at that site. As a result, we write

$$E^\circ_{cell} = E^\circ_{cathode} - E^\circ_{anode}$$
$$= (1.23 - x)V$$

We did not observe a reaction, so we might conclude that $1.23 - x$ is negative, or

$$1.23 - x < 0$$

and

$$x > 1.23\ V$$

This result is erroneous. As we have seen, our other observations have led us to the conclusion that $x < 0.80$ V.

This example suggests an important piece of advice: Be suspicious when you observe no reaction. Be especially suspicious if you are using a strong oxidizing agent or a strong reducing agent. If you suspect that a slow reaction rate is responsible, how should you proceed? One approach is to be patient and give the reaction more time to occur. Another approach is to increase the temperature. As a general rule, the rate of a reaction approximately doubles for each 10° increase in the temperature.

Concept of the Experiment

Table 19C.1 contains the half-cell reactions that pertain to this experiment, along with the corresponding electrode potentials. The unknown electrode potential for the VO_2 (*aq*), VO(*aq*), H_3O^+(*aq*) | Pt electrode is represented by *x*.

Table 19C.1

Standard Electrode Potentials at 25°C

Cathode Reaction	$E°$(V)
$VO_2^+ + 2H + e^- \rightleftharpoons VO^{2+} + H_2O$	*x*
$H_2O_2 + 2H_3O^+ + 2e^- \rightleftharpoons 4H_2O$	1.78
$Br_2 + 2e^- \rightleftharpoons 2Br^-$	1.07
$NO_3^- + 3H_3O^+ + 2e^- \rightleftharpoons HNO_2 + 4H_2O$	0.94
$SO_4^{2-} + 4H_3O^+ + 2e^- \rightleftharpoons H_2SO_3 + 5H_2O$	0.17
$2CO_2 + 2H_3O^+ + 2e^- \rightleftharpoons H_2C_2O_4 + 2H_2O$	-0.49

The oxidation numbers for vanadium in VO_2^+ and VO^{2+} are +5 and +4, respectively. Do you see how these numbers were obtained? If not, remember that the oxidation number of oxygen in each of these ions is –2 (Ebbing/ Gammon, Section 4.5). If you keep the oxidation numbers of vanadium in mind, you should be able to see that the half-reaction in the table that involves vanadium is a reduction.

All the substances to the left of the arrows in this table are potential oxidizing agents. All the substances to the right of the arrows are potential reducing agents. You will look for reactions between VO_2^+—a potential oxidizing agent—and potential reducing agents. These are H_2O, Br^-, HNO_2 (nitrous acid), H_2SO_3 (sulfurous acid), and $H_2C_2O_4$ (oxalic acid), as shown in Table 19C.1. You will also look for reactions between VO^{2+}—a potential reducing agent—and potential oxidizing agents. Because of the limited solubility of CO_2 in an acidic solution, the only potential oxidizing agents in Table 19C.1 that you can use are H_2O_2, Br_2, NO_3^-, and SO_4^{2-}. The successful reactions and the failures (after you are satisfied that they really are failures) can be used to estimate the desired electrode potential.

How will you know whether a reaction has occurred? The signal will be a color change. The color of VO_2^+ is yellow, whereas the color of VO^{2+} is blue. As a result, the successful reduction of VO_2^+ with a colorless reducing agent causes the color to change from yellow to blue. Similarly, the successful oxidation of VO^{2+} with a colorless oxidizing agent causes the color to change from blue to yellow.

A reaction mixture may turn green, however. If it does so, there are two possible reasons. The green color may be due either to the presence of roughly equal quantities of yellow VO_2^+ and blue VO^{2+} or to a combination of blue VO^{2+} and a yellow color from a reagent you have added. You will have an opportunity to explore the implications further in the Prelaboratory Assignment. Nevertheless, if you obtain a green color, be prepared to make some effort to discover which of these possibilities caused it. Some mental reflection and perhaps a little more experimental work will be necessary. Did you add the correct quantity of an oxidizing or reducing agent? If the reaction appears to be slow, did you have sufficient patience to observe the final outcome?

Additional Note

The reaction mixtures in this experiment will be acidic because you will be using a 0.05 *M* solution of VO_2^+ ions in 1 *M* H_2SO_4. The acidity plays a role that may not be obvious to you. During this experiment, you will add Na_2SO_3 (sodium sulfite) and $NaNO_2$ (sodium nitrite) to the solution of VO_2^+ ions. The acid in this solution will convert the sulfite and nitrite ions to the corresponding weak acids,H_2SO_3 and HNO_2. As a result, H_2SO_3 rather than SO_3^{2-}, and HNO_2 rather than NO_2^-, are given in Table 19C.1

You will also need to generate VO^{2+} ions in this experiment. You will prepare a solution of these ions by reducing VO_2^+ ions with a tenfold excess of sodium sulfite. After the reduction has occurred, you will remove the remainder of the reducing agent by heating the solution. Heating allows SO_2 (sulfur dioxide) to escape from the solution according to the reaction

$$H_2SO_3(aq) \xrightarrow{\Delta} H_2O(l) + SO_2(g)$$

Unfortunately, sulfur dioxide has a foul, suffocating odor. This substance must be handled carefully to ensure that its odor does not permeate your laboratory.

Procedure

Getting Started

1. Obtain 5 small test tubes. Mark each one with a number for identification.
2. Set up a boiling-water bath using a beaker of water, a ring stand, an iron ring, a wire gauze, and a laboratory burner. The bath should be placed in a hood if one is available. If not, use an inverted conical filter funnel connected by rubber tubing to a water aspirator. Clamp the funnel so that it is positioned directly over the beaker in which VO^{2+} will be prepared.

   CAUTION: Avoid burning your fingers. Do not touch the iron ring or wire gauze during heating.

3. During this experiment, you will use solutions of oxalic acid. Observe the following safety precaution:

   CAUTION: Wash your hands thoroughly after using solutions containing oxalic acid because they are poisonous. You should wear gloves in the laboratory.

Attempting Reduction of VO_2^+

1. Add 2 mL of 0.05 *M* VO_2^+ in 1 *M* H_2SO_4 to each of the 5 test tubes.
2. To the first test tube, add a small portion (about a quarter the size of a pea) of solid Na_2SO_3. Allow the test tube to stand for several minutes before placing it in the boiling-water bath. Record your observations.

   CAUTION: Sulfur dioxide has a foul, suffocating odor. Work

   under a hood if possible. If not, use the inverted conical filter funnel and water aspirator described earlier to suck away escaping SO_2.

3. Test the effect of 20 drops of water in the second test tube, 20 drops of 0.2 *M* NaBr in the third

test tube, 20 drops of 0.2 *M* $NaNO_2$ in the fourth test tube, and 20 drops of 0.2 *M* $H_2C_2O_4$ in the fifth test tube. Allow the test tubes to stand for about 3 min. If no reaction has occurred, place them in the boiling-water bath for at least 10 min. Record your observations.

4. Discard the solutions as directed by your laboratory instructor. Wash the test tubes and rinse them with distilled water.

Preparing VO_2^+

1. Measure 15 mL of the 0.05 *M* VO_2^+ solution, using a clean graduated cylinder. Pour the solution into a clean 150-mL beaker.
2. Use a balance to measure the desired mass of Na_2SO_3 for the reduction of the VO_2^+ ions. You calculated this quantity, which is

   10 times the stoichiometric amount, in the Prelaboratory Assignment. Record your measurements.
3. Add this substance to the solution of the VO_2^+ ions. Stir or swirl until all of the solid has dissolved. Allow the solution to stand for several minutes.
4. Replace the beaker of water from the water bath with this beaker. Heat the solution gently until all of the SO_2 has been expelled. Follow the same safety precautions for this step as described for Step 2 of the previous section.

Attempting Oxidation of VO_3^+

1. Add 2 mL of the solution of VO_2^+ to each of 4 test tubes.
2. Test the effect of 20 drops of 3% H_2O_2 in the first test tube, 40 drops of bromine water in the second test tube, 20 drops of 0.2 *M* $NaNO_3$ in the third test tube, and 20 drops of 0.2 *M* Na_2SO_4 in the fourth test tube. Allow the test tubes to stand for about 3 min. If no reaction has occurred, place them in the boiling water bath for at least 10 min. Record your observations.
3. Discard the solutions as directed. Wash the test tubes, and rinse them with distilled water.

CAUTION: Before you leave the laboratory, make sure that your gas outlet and those of your neighbors are closed.

Date ______________________ Student Name ______________________
Course/Section ______________________ Team Members ______________________
Instructor ______________________ ______________________

Estimating an Electrode Potential

Prelaboratory Assignment

1. a. What electrode potential will you estimate in this experiment?

   b. How will you make the estimate?

2. a. Calculate E°_{cell} for the following reactions in terms of x.

   $$2VO^{2+}(aq) + Br_2(aq) + 6H_2O(l) \rightarrow 2VO_2^{+}(aq) + 2Br^{-}(aq) + 4H_3O^{+}(aq)$$

   $$2VO_2^{+}(aq) + 2Br^{-}(aq) + 4H_3O^{+}(aq) \rightarrow 2VO^{2+}(aq) + Br_2(aq) + 6H_2O(l)$$

   b. If the first reaction occurs but the second does not, what can be said about the value of x?

3. What will you do if no immediate reaction occurs when you are attempting to reduce VO_2^{+} ions or oxidize VO^{2+} ions?

4. a. What is the color of VO_2^+ ions? What is the oxidation number of vanadium?

b. What is the color of VO^{2+} ions? What is the oxidation number of vanadium?

c. Attempt to assign an oxidation number to vanadium for each color in the following sequence of color changes that might occur during the reduction of VO_2^+ ions. Assume that the reducing agent and its oxidation product are colorless.

yellow $\rightarrow$ green $\rightarrow$ blue

5. a. How will you prepare a solution of VO^{2+} ions in this experiment?

b. Calculate the required mass of the reducing agent if a tenfold excess is to be used.

Student name: _______________ Course/Section: ________________ Date: ______________

c. Explain why you should use a tenfold excess of the reducing agent. Think carefully about your answer.

d. How will the remainder of the reducing agent be removed after the reduction has occurred? Write the balanced chemical equation for this reaction.

6. What special safety precautions must be observed during this experiment?

Date		Student Name	
Course/Section		Team Members	
Instructor			

Estimating an Electrode Potential

Results

1. *Attempting reduction of VO_2^+*

	Observations	Oxidation Number of V After Attempt
H_2O		
NaBr		
$NaNO_2$		
$H_2C_2O_4$		

2. *Preparing VO^{2+}*

Mass of container and Na_2SO_3 (g): ______________

Mass of empty container (g): ______________

Mass of Na_2SO_3 (g): ______________

3. *Attempting oxidation of VO^{2+}*

	Observations	Oxidation Number of V After Attempt
H_2O_2		
Br_2		
$NaNO_3$		
Na_2SO_4		

Questions

1. a. Which reagents caused the reduction of VO_2^+ ions?

Student name: ______________ Course/Section: ______________ Date: ______________

b. Give the complete reaction for each of these reagents with VO_2^+. Calculate $E°_{cell}$ for each reaction in terms of x.

c. What are the possible values of x in terms of these reactions?

2. a. Which reagents caused the oxidation of VO^{2+} ions?

b. Give the complete reaction for each of these reagents with VO^{2+}. Calculate $E°_{cell}$ for each reaction in terms of x.

c. What are the possible values of x in terms of these reactions?

3. a. The narrowest range for x is given by

$$\_\_\_\_\_\_\_\_\_\_ < x < \_\_\_\_\_\_\_\_\_\_ \text{ V}$$

b. As a crude estimate of x, obtain the mean value of these limits. This mean value should be used in the next question.

4. a. Using the crude estimate of x, calculate $E°_{cell}$ for the reaction of VO^{2+} with O_2 if

$$O_2(g) + 4HO^{+}(aq) + 4e^{-} \rightarrow 6H_2O(l) \qquad E° = 1.23 \text{ V}$$

b. Explain why the solution of VO^{2+} ions did not react readily with O_2 in the air.

5. During the experiment, you added 20 drops of water to a test tube containing VO_2^{+}. You wanted to know whether water would reduce that ion. Why was this test unnecessary?

20. Natural Radioactivity

Introduction

The spontaneous emission of particles and radiation from an atom's nucleus is known as *radioactivity* (Ebbing/Gammon, Chapter 20). Two sources of radioactivity are artificially produced isotopes and certain isotopes that occur naturally. This experiment deals with potassium-40, a naturally occurring radioactive isotope.

Purpose

In this experiment, you will measure the rate of disintegration of potassium-40 in a sample of ordinary KCl and calculate the half-life of this isotope.

A Special Note

Many people tend to associate radioactivity with events that they hope they will never experience, such as an accident at a nuclear power plant, cancer therapy, or even a nuclear explosion. However, all of us should be aware that we are always exposed to some radioactivity from natural sources. For example, every gram of potassium contains 0.00012 g of potassium-40, a radioactive isotope. Because potassium is an essential element in our bodies, all of us are slightly radioactive. Moreover, all of us are immersed in a sea of natural background radiation from several sources, including potassium-40 in the crust of the earth, radium and its decay products, and cosmic rays from space.

More Information About Potassium

Naturally occurring potassium consists of three isotopes. Only one, potassium-40, or $^{40}_{19}K$, is radioactive, and its abundance is only 0.012%.

This isotope undergoes radioactive decay by beta emission, electron capture, and positron emission (Ebbing/Gammon, Section 20.1). The frequency of positron emission is small compared with the other two modes of decay. The half-life of potassium-40 is 1.28×10^9 years.

Measuring the Radioactivity

You will use a Geiger counter (Ebbing/Gammon, Section 20.3) in this experiment. The units for the meter on this instrument will be counts/unit time. Each count represents the disintegration of a nucleus, so the counting rate that you measure represents the number of nuclei that are disintegrating in the unit of time. This is also called the *activity*.

The Geiger counter will not be completely efficient. As a result, you will not be able to count every disintegration of the potassium-40 nuclei. Some disintegrations will escape detection. There are at least three reasons for this. First, some of the radiation will be stopped by the sample itself before it reaches the Geiger counter. Second, not all the radiation that escapes from the sample will enter the counter. Third, the radiation that manages to enter the counter may not have enough energy to produce an electronic signal. Unless your laboratory instructor tells you differently, you can assume that one-fifth of the radiation from your sample has produced a signal.

Concept of the Experiment

First, you will measure the counting rate from background radiation. Then you will measure the counting rate due to the background and your sample of KCl. From these measurements, you will calculate the activity of potassium-40 in your sample that has been detected by the Geiger counter. You can then calculate the "true" activity from this result and the efficiency of the Geiger counter. In the

Prelaboratory Assignment, you will be able to show how these calculations will be carried out.

The half-life of potassium-40 can be calculated from the "true" activity of your sample, its measured mass, and the abundance of the isotope. The following steps are required. First, you will calculate the radioactive decay constant k for potassium-40. This constant will be obtained from the following equation (Ebbing/Gammon, Section 20.4):

$$\text{Rate} = kN_t$$

The rate is the "true activity," expressed in disintegrations per second (or nuclei undergoing disintegration per second), and N_t is the number of potassium-40 nuclei in your sample of KCl. Next, you will calculate the half-life $t_{1/2}$ from the following equation (Ebbing/Gammon, Section 20.4):

$$t_{1/2} = 0.693/k$$

Alternatively, you may wish to combine these equations and calculate the half-life directly from the equation

$$\text{Rate} = (0.693/t_{1/2})\ N_t$$

The half-life you have calculated may be far from the actual value. However, this method of determining the half-life of a long-lived nucleus is certainly more practical than actually waiting for half of the nuclei to decay. In the case of potassium-40, you and your descendants would have to wait slightly longer than a billion years.

Procedure

Getting Started

1. Your laboratory instructor may ask you to work with a partner.
2. Obtain a Geiger counter and instructions for using it.
3. Turn on your Geiger counter and let it warm up for 5 min. Remove the protective shield from the counter if one is present.

Doing the Experiment

1. Measure and record the background radiation by reading the meter 10 times at intervals of 10 s.

   Do not turn off the Geiger counter between readings.
2. Use a laboratory balance to measure the mass of a 150-mL beaker. Record your result.
3. Add 10-11 g of KCl to the beaker and measure the combined mass. Record your result, and calculate the mass of KCl in your sample.
4. As shown in Figure 20.1, clamp the counter as close to the sample in the beaker as possible without touching the sample.
5. Measure and record the count rate 10 times at intervals of 10 s. Do not turn off the Geiger counter between readings.

FIGURE 20.1

The arrangement of the Geiger counter within the beaker containing the sample of KCl.

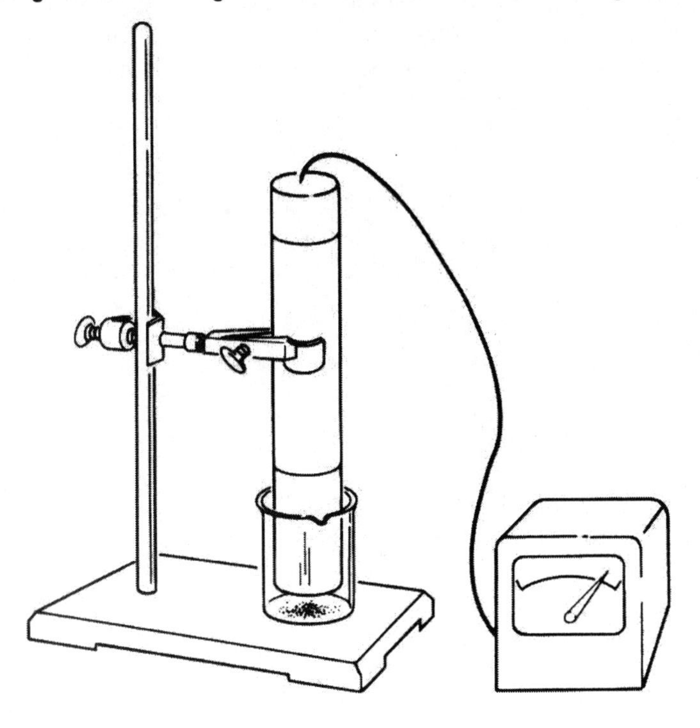

Date ______________________ Student Name ______________________
Course/Section ______________________ Team Members ______________________
Instructor ______________________ ______________________

Natural Radioactivity

Prelaboratory Assignment

1. Provide definitions for the following terms:

   a. Radioactivity

   b. Background radiation

   c. Isotope

   d. Beta emission

   e. Positron emission

   f. Electron capture

   g. Half-life

2. Write nuclear equations that describe the three modes of decay for potassium-40.

3. Write mathematical equations that show how you will calculate the detected activity from the count rate due to background radiation and the sample of KCl and how you will determine the "true" activity of your sample from the detected activity. Use the following symbols:

   R(background): count rate for background radiation

   R(total): total count rate

   R(detected): detected count rate from your sample

   R(true): "true" count rate of your sample

ate ______________ Student Name ______________
Course/Section ______________ Team Members ______________
nstructor ______________ ______________

Natural Radioactivity

Results

Background activity (count/s):

_______ _______ _______ _______ _______

_______ _______ _______ _______ _______

Mass of KCl and beaker (g): _______

Mass of beaker (g): _______

Mass of KCl (g): _______

Activity of background and KCl (count/s):

_______ _______ _______ _______ _______

_______ _______ _______ _______ _______

Questions

1. Compute a mean value and the standard deviation for each of the following quantities. You may calculate the standard deviation using a computer and the Internet or by hand. The procedure for calculating a standard deviation are discussed in Appendix A. Give, for each quantity, a mean value that is consistent with the precision of the data.

    a. The background activity:

    b. The total activity:

tudent name: _______________ Course/Section: _________________ Date: _______________

c. Calculate the "true" activity of potassium-40 in your sample of KCl. Your answer should be consistent with the precision of the data used in the calculations.

d. Calculate the half-life of potassium-40, and compare your value with the known value.

2. The average human body has 1.4×10^2 g of potassium within it. Calculate the activity in disintegrations/year of the potassium-40 in the average human body, using your value for the half- life of this substance.

21A. Qualitative Analysis of Mg^{2+}, Ca^{2+}, Ba^{2+}, and Al^{3+} Ions

Introduction

Some of the similarities and differences in the chemical behaviors of the first three main-group families (Ebbing/Gammon, Chapter 21) can become very clear during qualitative analysis by chemical methods.

Purpose

You will learn to separate and identify each cation in a mixture of Mg^{2+}, Ca^{2+}, Ba^{2+}, and Al^{3+} ions. Your studies will enable you to determine which of these ions are present in an unknown mixture. This mixture may contain any or all of these cations.

Concept of the Experiment

The characteristic chemical behaviors of Mg^{2+}, Ca^{2+}, Ba^{2+}, and Al^{3+} ions will become clearer to you as you learn to separate and identify each cation. You will discover different chemical behaviors within a family by comparing the behaviors of Mg^{2+}, Ca^{2+}, and Ba^{2+} ions. You will discover differences between cations in different families but in the same period from the behaviors of Mg^{2+} and Al^{3+} ions. There are four principal tasks in this experiment:

1. The separation of the cation from Group 3 (Al^{3+}) from the cations belonging to Group 2 (Mg^{2+}, Ca^{2+}, and Ba^{2+})
2. The identification of Al^{3+} by characteristic chemical reactions
3. The separation of the cations in Analytical Group 2
4. The identification of these cations by characteristic chemical reactions

Each of these tasks will be discussed in turn.

The Separation of Groups 2 and 3

This separation will be brought about by the precipitation of $Al(OH)_3$. The precipitation must be selective, because $Mg(OH)_2$ is also essentially insoluble and $Ca(OH)_2$ is only slightly soluble. The solubility product constants for these substances are given in Table 21.1.

Table 21.1

Solubility Product Constants

Compound Formula	Compound Name	K_{sp}
$Al(OH)_3$	Aluminum hydroxide	4.6×10^{-23}
$Mg(OH)_2$	Magnesium hydroxide	1.8×10^{-11}
$Ca(OH)_2$	Calcium hydroxide	5.5×10^{-6}
$Ba(OH)_2$	Barium hydroxide	Soluble
$MgCrO_4$	Magnesium chromate	Soluble
$CaCrO_4$	Calcium chromate	7.1×10^{-4}
$BaCrO_4$	Barium chromate	1.2×10^{-10}
MgC_2O_4	Magnesium oxalate	8.5×10^{-5}
CaC_2O_4	Calcium oxalate	2.3×10^{-9}
BaC_2O_4	Barium oxalate	1.6×10^{-7}

You will achieve selective precipitation by using a buffered solution of NH_3, a weak base, rather than a solution of a strong base, such as NaOH. The buffer (Ebbing/Gammon, Section 16.6) will fix the hydroxide ion concentration so that only $Al(OH)_3$, the most insoluble hydroxide in Table 21.1, will precipitate.

To prepare the buffer, a *limited* amount of HCl is added to an *excess* quantity of NH_3. Because the reaction of these substances yields NH_4Cl, a buffer containing NH_3 and the ammonium ion, NH_4^+,will result. The hydroxide ion concentration is fixed at a desired value by the relative molar quantities of the base and its conjugate acid.

The amphoteric nature of $Al(OH)_3$ (Ebbing/Gammon, Section 21.4) provides another reason for using the buffer solution. If a strong base were used, $Al(OH)_3$ would precipitate, as you would expect. However, this precipitate would dissolve in the presence of any excess hydroxide ions because of the reaction

$$Al(OH)_3(s) + OH^-(aq) \rightleftharpoons Al(OH)_4^-(aq)$$

The characteristic reactions that are used to provide positive identification of Al^{3+} are based on the amphoteric nature of $Al(OH)_3$ and on the insolubility of this substance in a solution of the weak base NH_3.

The Separation of the Cations in Group 2

The chromates and oxalates of Mg^{2+}, Ca^{2+}, and Ba^{2+} play important roles in the separation of these cations. The solubility product constants for these compounds are given in Table 21.1.

You will notice that the solubilities of the chromates decrease from Mg^{2+} through Ba^{2+}. The solubilities of these compounds are influenced by the acidity of the solution because of the equilibrium between yellow CrO_4^{2-} (chromate ion) and orange $Cr_2O_7^{2-}$ (dichromate ion):

$$2CrO_4^{2-}(aq) + 2H_3O^+(aq) \rightleftharpoons Cr_2O_7^{2-}(aq) + 3H_2O(l)$$

Nevertheless, K_{sp} for $BaCrO_4$ is sufficiently small that this substance precipitates even from a weakly acidic solution. Precipitation of $MgCrO_4$ and $CaCrO_4$ does not occur under this condition.

There is no regular trend in the solubilities of the oxalates, but CaC_2O_4 is the most insoluble of these substances. Oxalic acid is a weak diprotic acid, so an equilibrium between $C_2O_4^{2-}$ and $HC_2O_4^-$ exists in a weakly acidic solution:

$$C_2O_4^{2-}(aq) + H_3O^+(aq) \rightleftharpoons HC_2O_4^-(aq) + H_2O(l)$$

Consequently, the acidity of the solution again affects the solubilities. Unlike MgC_2O_4 and BaC_2O_4, CaC_2O_4 precipitates from a weakly acidic solution because K_{sp} for this compound is very small.

Now consider a weakly acidic solution of Mg^{2+}, Ca^{2+}, and Ba^{2+}. The addition of K_2CrO_4 will cause $BaCrO_4$ to precipitate and allow the separation of Ba^{2+} from Mg^{2+} and Ca^{2+}. Subsequently, the addition of $K_2C_2O_4$ will result in the precipitation of CaC_2O_4 and allow the separation of Ca^{2+} from Mg^{2+}.

The Identification of Mg^{2+}, Ca^{2+}, and Ba^{2+}

After $BaCrO_4$ is dissolved in a solution containing a strong acid, Ba^{2+} will be precipitated once again as white $BaSO_4$ ($K_{sp} = 1.1 \times 10^{-10}$) by the addition of Na_2SO_4. This reaction will confirm the presence of Ba^{2+}.

A strong acid will be used to dissolve CaC_2O_4. This substance will be precipitated again by making the solution basic. This reaction will confirm the presence of Ca^{2+}.

Finally, Mg^{2+} will be precipitated as $Mg(OH)_2$. Although this compound is usually colorless when it is freshly precipitated, it will be stained yellow in this experiment because of the presence of yellow CrO_4^{2-}. This precipitate will be dissolved in an acidic solution. White $MgNH_4PO_4$ ($K_{sp} = 2.5 \times 10^{-12}$) will then be precipitated. Both of these will be used to confirm the presence of Mg^{2+}.

Procedure

Getting Started

1. Obtain 6 small test tubes.
2. Obtain your unknown mixture and record its identification number.
3. Obtain 1 mL of the known mixture. This solution contains $Mg(NO_3)_2$ (0.2 *M*), $Ca(NO_3)_2$ (0.2 *M*), $Ba(NO_3)_2$ (0.2 *M*), and $Al(NO_3)_3$ (0.2 *M*).

   CAUTION: Wash your hands thoroughly after using the solution containing barium, because it is poisonous. You may wish to wear gloves if they are available in the laboratory.

4. Conduct the analysis of the known and unknown solutions simultaneously so that you can compare the results.
5. Use labeled test tubes throughout the experiment so that you do not confuse the known and unknown solutions and precipitates at any time.
6. If necessary, obtain instructions from your laboratory instructor for using the centrifuges in your laboratory.

   CAUTION: When you use a centrifuge, do not attempt to stop the centrifuge rotor with your finger or anything else.

7. Obtain directions from your laboratory instructor for discarding the solutions that you will use in this experiment.

8. Take care in handling the solutions used in this experiment.

CAUTION: The 6 *M* solutions of ammonia, hydrochloric acid, and sodium hydroxide can cause chemical burns, in addition to ruining your clothes. If you spill any of these on you, wash the contaminated area thoroughly with water and immediately report the incident to your instructor. You may require further treatment.

Doing the Analysis

1. Take 1 mL of the known mixture and 1 mL of the unknown mixture in separate small test tubes.

2. Each of the subsequent additions and operations should be conducted on both the known and the unknown mixtures unless the instructions indicate otherwise.

3. Add 6 drops of 6 *M* NH_3 and 4 drops of 6 *M* HCl to each test tube to make buffer solutions. Look carefully for a precipitate because sometimes it is difficult to see. If a clear solution becomes cloudy, it is likely that a precipitate has been formed.

4. If no precipitate forms, proceed with Step 8. If a precipitate forms, stir the mixture with a clean stirring rod. Centrifuge the mixture for about 1 min.

5. Decant (pour off) the solution into a clean test tube. Save this solution for Step 8. Use the precipitate in the following step.

6. Wash the precipitate by adding 1 mL of distilled water, 6 drops of 6 *M* NH_3, and 4 drops of 6 *M* HCl and stirring the mixture vigorously. Centrifuge the mixture, save the precipitate, and discard the solution.

7. Dissolve the precipitate by gently shaking it with 4 drops of 6 *M* NaOH. Add drops of 6 *M* HCl until a drop of the solution transfered using a clean stirring rod to a piece of blue litmus paper turns the paper pink. A white precipitate may appear briefly during this operation. Add drops of 6 *M* NH_3 until a white precipitate appears. These reactions confirm the presence of Al^{3+}.

8. Add 1 *M* acetic acid by drops to the solution from either Step 4 or Step 5 until a drop of the solution transferred using a clean stirring rod to a piece of blue litmus paper turns the paper pink. Avoid using excess acetic acid.

9. Add 8 drops of 1 *M* K_2CrO_4 to the solution, and stir the mixture thoroughly with a clean stirring rod.

10. If no precipitate has formed, proceed with Step 13. If a yellow precipitate has formed, centrifuge the mixture, save the solution for Step 13, and use the precipitate in the next step.

11. Dissolve the precipitate in 3 drops of 6 *M* HCl and 3 drops of distilled water.

12. Add 20 drops of 0.1 *M* Na_2SO_4. A white precipitate in a yellow solution will result. To show that the precipitate is white, centrifuge the mixture and discard the solution. You may wash the precipitate with distilled water if desired. The formation of this precipitate confirms the presence of Ba^{2+}.

13. Add 10 drops of 1 *M* $K_2C_2O_4$ to the solution from Step 10.

14. If a white precipitate does not appear within 10 min, proceed with Step 17. If a precipitate does appear, go on to the next step.

15. Stir the mixture with a clean stirring rod. Centrifuge the mixture and decant the solution. Save the solution for Step 17 and use the precipitate in the following step.

16. Dissolve the precipitate in 4 drops of 6 *M* HCl and add 20 drops of distilled water. Add another drop of 1 *M* $K_2C_2O_4$, and make the solution basic to litmus paper with 6 *M* NaOH. A white precipitate confirms the presence of Ca^{2+}.
17. Add 6 drops of 6 *M* NaOH to the solution from either Step 14 or Step 15. Stir and then centrifuge the mixture. Examine the mixture carefully. If no precipitate has formed, you have completed the analysis. If a translucent, pale yellow precipitate has formed, decant and discard the solution, and use the precipitate in the next step.
18. Dissolve the precipitate by adding 3 drops of 6 *M* HCl.
19. Add drops of 6 *M* NH_3 until the solution is basic to litmus.
20. Add 10 drops of 1 *M* Na_2HPO_4. A white precipitate, which may form slowly, confirms the presence of Mg^{2+}.
21. Record the cations that are present in the unknown mixture.

CAUTION: Wash your hands thoroughly. Oxalate solutions are poisonous. You may wish to wear gloves if they are available in the laboratory.

Date ______ Student Name ______
Course/Section ______ Team Members ______
Instructor ______

Qualitative Analysis of Mg^{2+}, Ca^{2+}, Ba^{2+}, and Al^{3+} Ions

Prelaboratory Assignment

1. Complete the following flow diagrams by filling in the boxes with the correct cations or precipitates.

   a. Initial separation of cations:

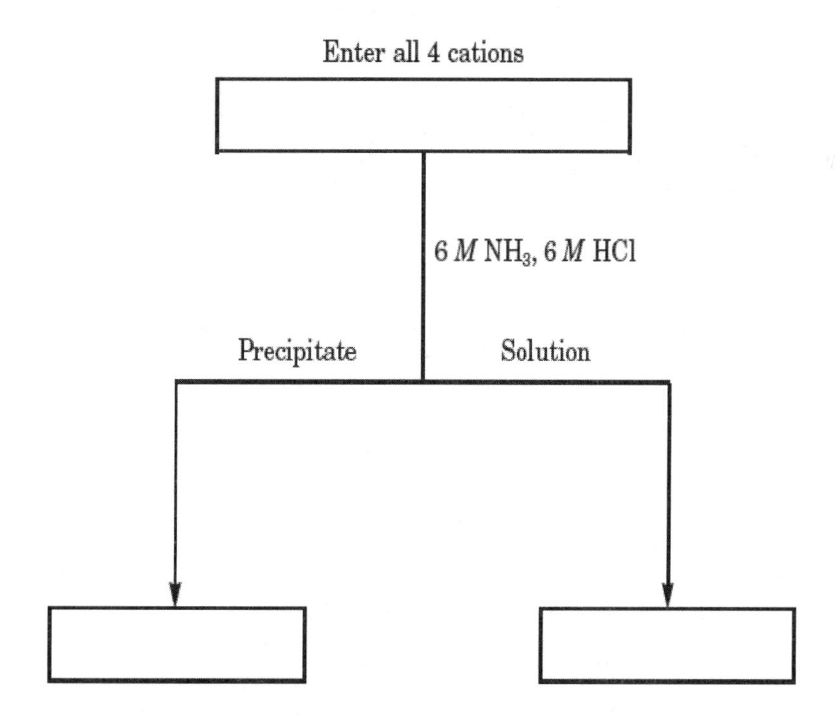

   b. Identification of cation from Group 3:

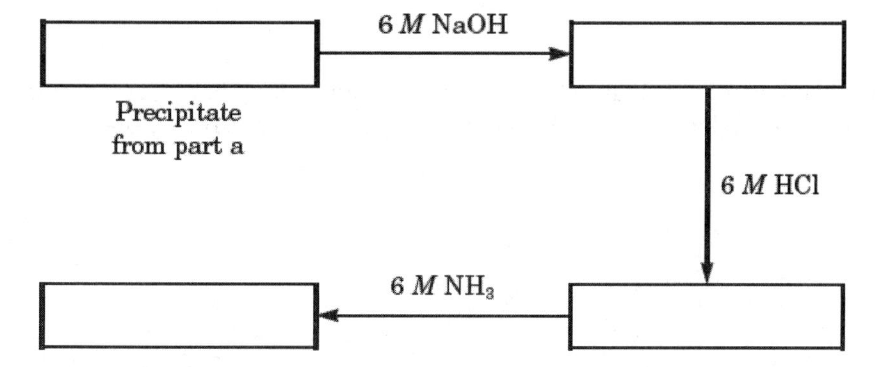

c. Separation of cations from Group 2:

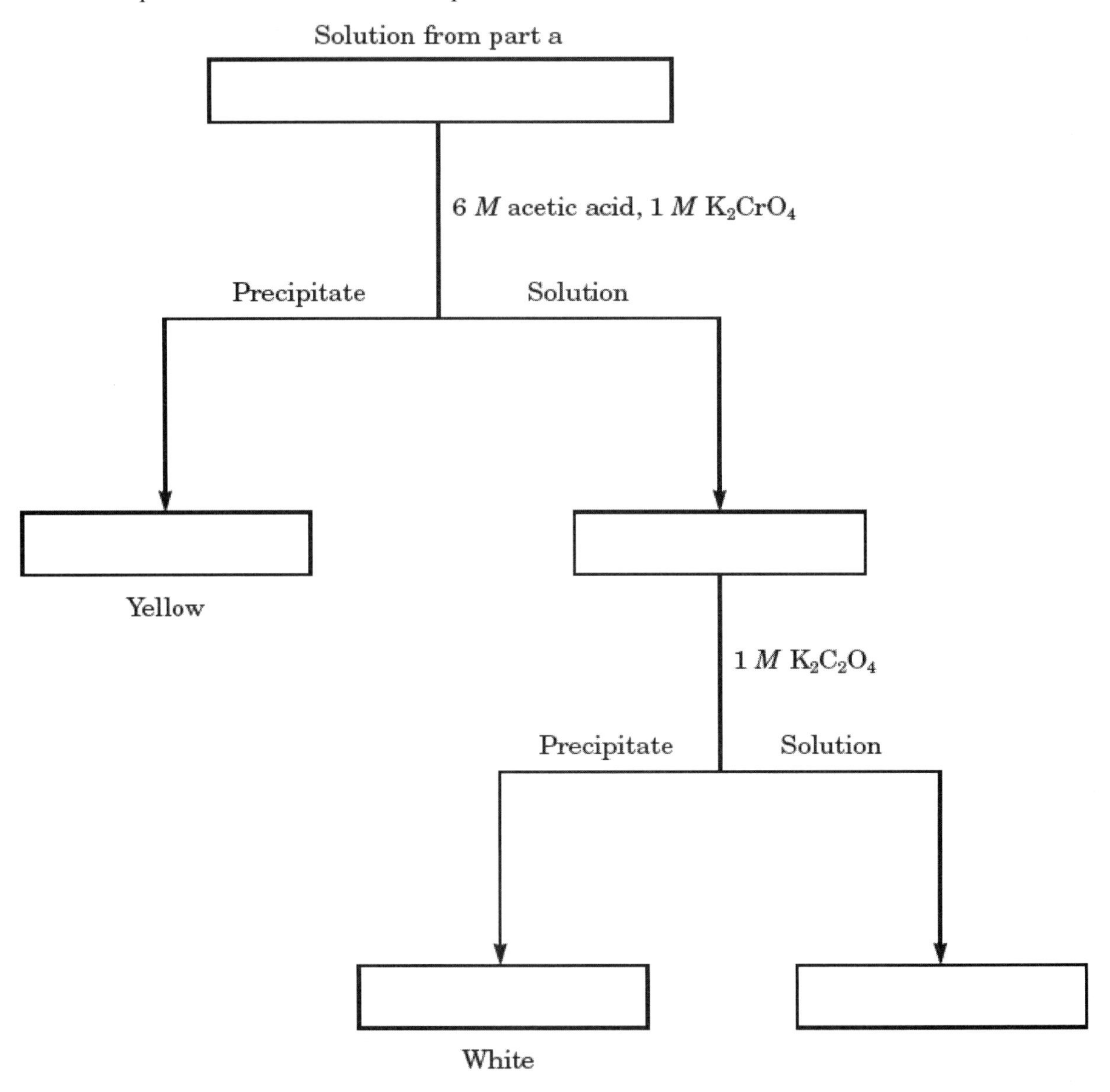

d. Devise a flowchart similar to those in parts a, b and c that shows the identification of the cation in the yellow precipitate from part c. Give the colors of any precipitates.

Student name: ________________ Course/Section: ________________ Date: ________________

e. Devise a flowchart that shows the identification of the cation in the white precipitate from part c. Give the colors of any precipitates.

f. Devise a flowchart that shows the identification of the remaining cation from part c. Give the colors of any precipitates.

2. Describe the purpose of the buffer solution used in this experiment.

3. a. What is an amphoteric substance?

b. What is the only substance in this experiment that is amphoteric?

4. What special safety precautions are cited in this experiment?

Date ____________________ Student Name ____________________

Course/Section ____________________ Team Members ____________________

Instructor ____________________ ____________________

Qualitative Analysis of Mg^{2+}, Ca^{2+}, Ba^{2+}, and Al^{3+} Ions

Results

Unknown no.: ________________

Ions present: ________________

Questions

1. a. Write a balanced equation for each precipitation reaction that was used to separate the four cations.

   b. Write a balanced equation for each reaction that was used to confirm the identity of each cation.

2. a. Suppose a mixture contained only Al^{3+} and Mg^{2+}. How would you *separate* these cations?

b. Suppose a mixture contained only Al^{3+} and Ba^{2+}. An unbuffered solution of NH_3 will allow these cations to be separated. Why?

c. Suppose a mixture contained only Mg^{2+} and Ba^{2+}. What would happen upon the addition of 1 *M* NaOH? 1 *M* K_2CrO_4? 1 *M* $K_2C_2O_4$? Use Table 21.1.

21B. The Strength of a Laundry Bleach

Introduction

Hypochlorous acid (HClO) is one of the important chlorine oxoacids (Ebbing/Gammon, Section 21.9). Solutions of *sodium hypochlorite* (NaOCl), a salt of that acid, are sold as laundry bleach. The hypochlorite anion (ClO^-) is a strong oxidizing agent, but not as strong as ClO_2^-, ClO_3^-, and ClO_4^-.

Purpose

In this experiment, you will use an oxidation–reduction titration to determine the quantity of NaOCl in a commercial bleach.

Concept of the Experiment

As you will see, you can determine the quantity of the ClO^- ion in a solution through two oxidation–reduction reactions. First, a known quantity of this anion is reduced to Cl^- ions in an acidic solution, using excess potassium iodide. The I^- ions are oxidized to I_2 in this reaction. The solution that results is brown because that is the color of I_2 in water. Second, the I_2 is reduced to I^- during a titration with asolution of sodium thiosulfate. Thiosulfate anions ($S_2O_3^{2-}$) are oxidized to tetrathionate ions ($S_4O_6^{2-}$) in this reaction.

You may have found this description somewhat confusing, but it should become clear after you have balanced the equations for the two oxidation–reduction reactions in the Prelaboratory Assignment. Moreover, the relationship between the original quantity of ClO^- ions and the quantity of $S_2O_3^{2-}$ ions used in the titration will be evident after these equations are balanced.
Although you could use the disappearance of the color due to aqueous I_2 to detect the endpoint of the titration, this technique would not be very sensitive. Instead, you will use starch as an indicator. Starch reacts with I_2 to form a dark blue color. This reaction is reversible. Consequently, the blue color fades during the course of the titration as I_2 is consumed. The endpoint occurs when *one* drop of the $Na_2S_2O_3$ solution causes the color to change from blue to colorless. A trial titration will enable you to locate the endpoints of subsequent titrations more easily.

Procedure

Getting Started

1. Obtain a 10-mL transfer pipet and a 50-mL buret.
2. Obtain about 70 mL of a 0.0250 *M* solution of $Na_2S_2O_3$ and 3 samples of solid KI. Each of these samples should have a volume of about 1 cm_3. They can be stored on pieces of waxed paper.
3. A buret containing the bleach solution will be available for general use. This solution must be diluted, however. Record the initial buret reading to the nearest 0.01 mL. Carefully add about 3 mL of the solution to a 100-mL volumetric flask. (A 100-mL graduated cylinder may be used if a volumetric flask is not available.) Record the final buret reading to the nearest 0.01 mL, and calculate the volume of undiluted bleach used. Add distilled water to the flask until the bottom of the meniscus coincides with the etched line on the flask. Add the last 0.5 mL with a medicine dropper. Insert a stopper in the flask, and mix the solution thoroughly.
4. Be careful in your handling of the solutions used in this experiment. Obtain directions from your laboratory instructor regarding proper disposal of all solutions used during this experiment.

CAUTION: Undiluted bleach and hydrochloric acid can cause chemical burns and ruin your clothes. In addition, bleach can be especially irritating to the eyes. If you spill either of these solutions on you, wash the contaminated area thoroughly and immediately report the incident to your laboratory instructor. You may require further treatment. Be sure to wear approved chemical splash goggles.

Cleaning and Filling your Buret

1. Instructions for using a buret can be found in the Introduction section of this manual. Clean your buret according to the directions, and fill it with some of the $Na_2S_2O_3$ solution.

Doing the Trial Titration

1. Pipet 10.0 mL of the diluted bleach solution into a clean 125-mL Erlenmeyer flask.
2. Add one of the samples of solid KI to the flask, followed by 20 mL of distilled water from a graduated cylinder and 20 drops of 2 *M* HCl. Swirl to obtain a homogeneous solution.
3. Record the initial buret reading to the closest 0.01 mL.
4. Place the flask under the buret with the capillary tip inside the mouth of the flask. Insert a piece of white paper under the flask.
5. While swirling, add increments of about 1 mL of the $Na_2S_2O_3$ solution to the flask. Continue until the brown color fades to yellow.
6. Add 40 drops of a 0.2% starch solution to the flask.
7. Continue to add 1-mL portions of the $Na_2S_2O_3$ solution until one addition causes the solution to become colorless.
8. Record the final buret reading to the nearest 0.01 mL, and calculate the approximate volume of the $Na_2S_2O_3$ solution required for the titration.

Doing the Exact Titrations

1. Repeat Steps 1 through 4 of the procedures for the trial titration.
2. Subtract 2 mL from the volume found in the trial titration. Rapidly add this new volume to the flask from the buret.
3. Add 40 drops of the 0.2% starch indicator to the flask.
4. Rinse the walls of the flask with distilled water from a plastic wash bottle.
5. Continue the titration on a *drop-by-drop* basis. Swirl the flask rapidly after each addition.

   Remember that the endpoint will occur when one drop results in a colorless solution.
6. Record the final buret reading to the nearest 0.01 mL.
7. Repeat the procedure with a second sample of the diluted bleach solution.
8. If the volumes at the endpoints differ by more than 0.15 mL (about 3 drops) or some other amount specified by your instructor, repeat the titrations with additional samples of the diluted bleach until the required precision is obtained.
9. Calculate and record the molarity of the diluted bleach solution. Obtain the mean molarity.

Date	______________	Student Name	______________
Course/Section	______________	Team Members	______________
Instructor	______________		______________

The Strength of a Laundry Bleach

Prelaboratory Assignment

1. Write the balanced equations for the following oxidation–reduction reactions that you will encounter in this experiment:

   a. The reduction of ClO^- by I^- in acidic solution

   b. The reduction of I_2 by $S_2O_3^{2-}$ in acidic solution

2. What conversion factor will enable you to calculate the number of moles of ClO^- ions from the number of moles of $S_2O_3^{2-}$ ions used in the titration?

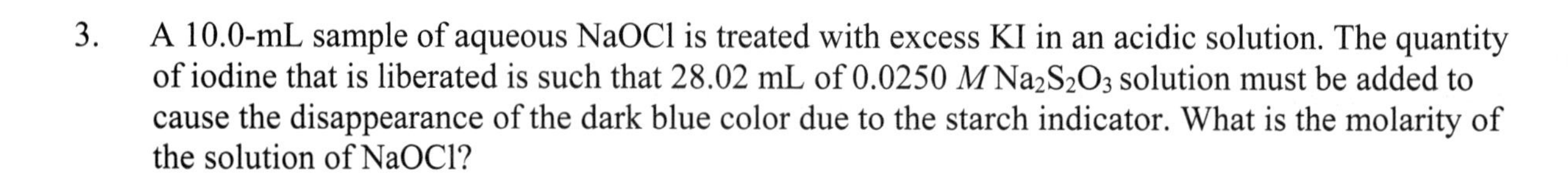

3. A 10.0-mL sample of aqueous NaOCl is treated with excess KI in an acidic solution. The quantity of iodine that is liberated is such that 28.02 mL of 0.0250 *M* $Na_2S_2O_3$ solution must be added to cause the disappearance of the dark blue color due to the starch indicator. What is the molarity of the solution of NaOCl?

4. What safety precaution is cited in this experiment?

Date	______________	Student Name	______________
Course/Section	______________	Team Members	______________
Instructor	______________		______________

The Strength of a Laundry Bleach

Results

1. *Quantitative dilution of the bleach*

   Final buret reading (mL): ______________

   Initial buret reading (mL): ______________

   Volume of undiluted bleach (mL): ______________

2. *Trial titration*

   Volume of diluted bleach used (mL): ______________

   Final buret reading (mL): ______________

   Initial buret reading (mL): ______________

   Volume of 0.0250 M $Na_2S_2O_3$ (mL): ______________

3. *Exact titrations*

Trial	1	2	3	4
Volume of diluted bleach used (mL)	________	________	________	________
Final buret reading (mL)	________	________	________	________
Initial buret reading (mL)	________	________	________	________
Volume of 0.0250 M $Na_2S_2O_3$ (mL)	________	________	________	________
Moles of $Na_2S_2O_3$	________	________	________	________
Moles of NaOCl	________	________	________	________
Molarity of NaOCl in diluted bleach (M)	________	________	________	________

Mean molarity (M) ________

Calculations:

Student name: ________________ Course/Section: __________________ Date: ________________

Questions

1. a. Calculate the molarity of the undiluted bleach solution.

   b. Commercial bleaches such as Clorox are said to contain 5.25% NaOCl by weight. Is this correct? Use your results and a density of 1.0 g/mL for the bleach in your calculations. These calculations should reflect the precision of the assumed density.

2. Redo the third problem of the Prelaboratory Assignment, substituting $NaIO_3$ (sodium iodate) for NaOCl. The new reaction yields I_2 as the only product containing a halogen.

22. Qualitative Analysis of Chromium, Iron, and Copper

Introduction

We have used copper and iron as basic materials since the Bronze and Iron Ages, but our extensive use of chromium began only after the Industrial Revolution. Some of the similarities and differences in the chemical behaviors of the ions of these materials can become very clear during *qualitative analysis* by chemical methods. Qualitative analysis is the identification of the substances in a mixture. When chemical methods are used in the identification of mixtures of metal cations, these ions are usually separated before identification can occur. After they are separated, identification of each cation depends on the observation of a characteristic chemical reaction.

Purpose

You will learn to separate mixtures of Cr^{3+}, Fe^{3+}, and Cu^{2+} and to identify each ion. Finally, you will be able to determine which of these ions are present in an unknown mixture.

Concept of the Experiment

Separation begins when aqueous ammonia is added to a mixture of the three ions. Because aqueous ammonia is a weak base, it is a source of ammonia as well as a source of hydroxide ions. As a consequence, this reagent causes the precipitation of the hydroxides of Cr^{3+} and Fe^{3+}. It also causes the formation of $Cu(NH_3)_4^{2+}$, a deep blue complex ion, which remains in solution.

Decanting (pouring off) the solution from the precipitate separates copper from chromium and iron. The presence of copper is confirmed when a red-maroon precipitate of $Cu_2Fe(CN)_6$ forms after the addition of an acid (to neutralize NH_3) and a solution of $K_4Fe(CN)_6$.

To continue the separation, the hydroxides of Cr^{3+} and Fe^{3+} are treated with hydrogen peroxide in basic solution. Only $Cr(OH)^3$ is oxidized. Yellow CrO_4^{2-} (chromate ion) is formed in solution, whereas $Fe(OH)_3$ remains as a precipitate. When the solution is decanted from the precipitate, chromium and iron separate. The presence of chromium in the solution is confirmed when a yellow precipitate of $PbCrO^4$ forms after a solution of $Pb(NO_3)_2$ is added.

The precipitate of $Fe(OH)_3$ is then dissolved in an acid. The presence of iron is confirmed by the formation of a deep red color due to $Fe(SCN)^{2+}$, a complex ion, after the addition of a solution of KSCN.

In the Prelaboratory Assignment you will construct a flowchart for the entire separation scheme in order to become more familiar with it. You can use this scheme to separate and identify any combination of these metal ions in an unknown mixture. Also note that each of these metal ions has a characteristic color. As a result, you should use the color of your unknown mixture to support your analysis using the scheme.

Procedure

Getting Started

1. Make sure you are familiar with the flowchart that you completed in the Prelaboratory Assignment.
2. Obtain 6 small test tubes.
3. Obtain your unknown solution, and record its identification number and color.
4. Prepare a known mixture of the three ions, using 5 mL of a 0.1 *M* solution of $Cr(NO_3)_3$, 5 mL of a 0.1 *M* solution of $Fe(NO_3)_3$, and 5 mL of a 0.1 *M* solution of $Cu(NO_3)_2$. Note and record the color of each of these solutions before you add it. Make sure that the final solution is thoroughly mixed.
5. Conduct the analysis of the known and the unknown solutions simultaneously so that you can compare the results.
6. Use labeled test tubes so that you do not confuse the known and unknown solutions and precipitates at any time.
7. If necessary, obtain instructions from your laboratory instructor for using the centrifuges in your laboratory.

   CAUTION: When you use a centrifuge, do not attempt to stop the centrifuge rotor with your finger or anything else.

8. Obtain directions from your laboratory instructor for discarding the solutions that you will use in this experiment.
9. Be careful handling the solutions used in this experiment.

   CAUTION: Ammonia, sodium hydroxide, nitric acid, and acetic acid can cause chemical burns, in addition to ruining your clothes. If you spill any of these on you, wash the contaminated area thoroughly with tap water, and immediately report the incident to your instructor. You may require further treatment.

Doing the Analysis

1. Take 1 mL of the known mixture and 1 mL of the unknown mixture in separate test tubes.
2. Each of the following additions and operations should be conducted on both mixtures.
3. Add 20 drops of 6 *M* NH_3 and stir with a clean stirring rod.
4. If no precipitate forms, proceed with Step 12. If a precipitate is present, centrifuge the mixture for about 1 min and decant the solution. Save the solution for Step 12 and use the precipitate in Step 5.
5. Spray the precipitate with about 1 mL of distilled water from a plastic wash bottle. Stir with a clean stirring rod, and then centrifuge the mixture. Discard the water in the appropriate waste container and use the precipitate in the following step.
6. Add 12 drops of 3% H_2O_2, 1 mL of distilled water, and 5 drops of 6 *M* NaOH to the precipitate. Let this mixture stand for about 1 min.

7. Light a laboratory burner, and heat the mixture gently and cautiously to decompose the remaining H_2O_2. Use a test tube holder. Do not let the flame linger in any one place, or "bumping" will occur. Heat until effervescence (bubbling) from the decomposition ceases.

8. Cool the test tube briefly under tap water. Centrifuge the mixture. Save the solution in a clean test tube for Step 13. If a precipitate remains, use it in Step 9. If no precipitate remains, proceed with Step 13.

9. Wash the precipitate in the same manner as in Step 5.

10. Dissolve the precipitate with 5 drops of 6 *M* HNO_3. Add 3 mL of distilled water, and mix to obtain a homogeneous solution.

11. Add 5 drops of 0.1 *M* KSCN to this solution, and mix thoroughly. A deep-red color confirms the presence of iron.

12. Make the solution from Step 4 acidic by adding drops of 6 *M* acetic acid until the solution turns blue litmus paper pink. Add 10 drops of 0.1 *M* $K_4Fe(CN)_6$ and mix thoroughly. A red-maroon precipitate confirms the presence of copper.

13. If you did Step 8, make the solution acidic by adding drops of 6 *M* acetic acid until the solution turns blue litmus paper pink. Add 10 drops of 0.1 *M* $Pb(NO_3)_2$ and mix thoroughly. Centrifuge the mixture. A yellow precipitate confirms the presence of chromium.

14. Record the ions that are present in the unknown mixture. Is the color of the unknown mixture in agreement with your conclusions? If not, you will need to repeat the analysis.

CAUTION: Before you leave the laboratory, make sure that your gas outlet and those of your neighbors are closed. Finally, wash your hands. Solutions containing lead are poisonous. You may wish to wear gloves if they are available in the laboratory.

Date ______________________ Student Name ______________________

Course/Section ______________________ Team Members ______________________

Instructor ______________________ ______________________

Qualitative Analysis of Chromium, Iron, and Copper

Prelaboratory Assignment

1. Complete the following flowchart by inserting the reagents and products from each reaction.

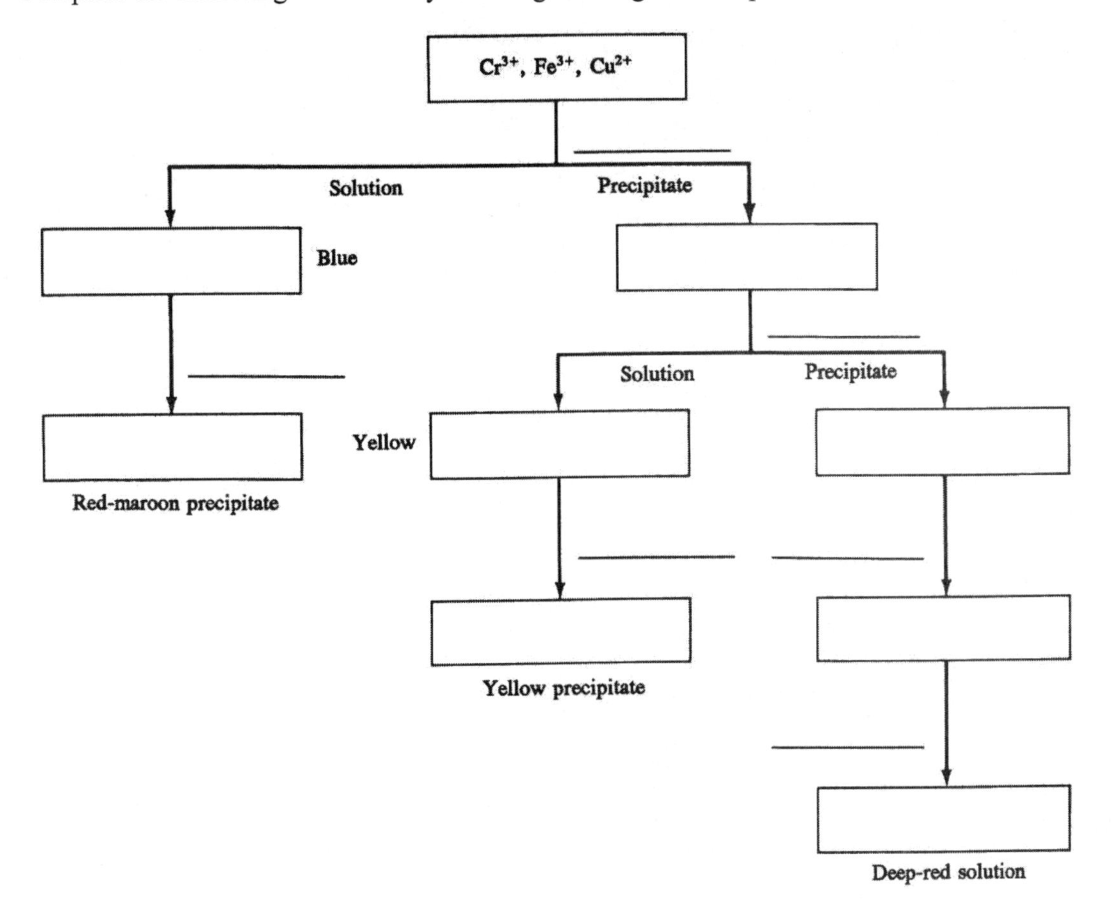

2. What special safety precautions are cited in this experiment?

Date	______________________	Student Name	______________________
Course/Section	______________________	Team Members	______________________
Instructor	______________________		______________________

Qualitative Analysis of Chromium, Iron, and Copper

Results

Unknown no.: ______________

Color of unknown mixture: ______________

Color of $Cr(NO_3)_3$ solution: ______________

Color of $Fe(NO_3)_3$ solution: ______________

Color of $Cu(NO_3)_2$ solution: ______________

Ions present in unknown mixture: ______________

Questions

1. Write a balanced chemical equation for each reaction that appears in the flowchart in the Prelaboratory Assignment.

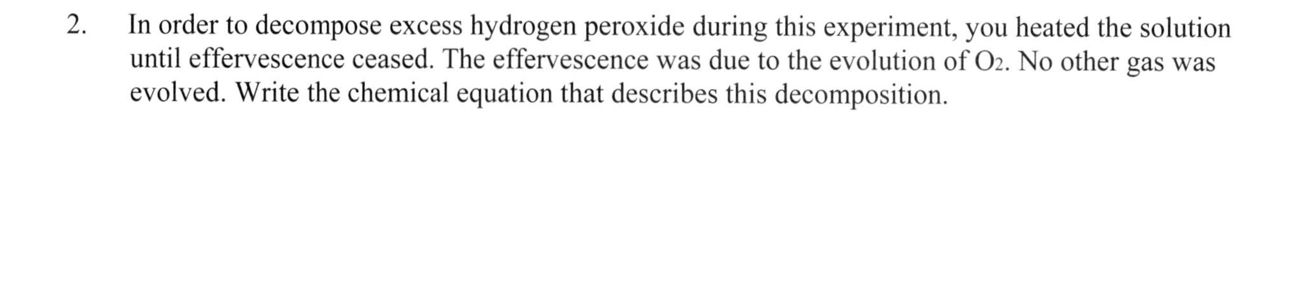

2. In order to decompose excess hydrogen peroxide during this experiment, you heated the solution until effervescence ceased. The effervescence was due to the evolution of O_2. No other gas was evolved. Write the chemical equation that describes this decomposition.

3. Suppose a solution contained only one cation: Cr^{3+}, Fe^{3+}, or Cu^{2+}. What would you do to identify the cation quickly?

23. Molecular Models of Organic Molecules

Introduction

Organic chemistry is the chemistry of compounds that contain carbon. Several million organic compounds are now known. The large number of these compounds is due to the carbon atom's ability to bond to other carbon atoms as well as to atoms of many other elements. The diversity of the structural and geometric arrangements in organic compounds is one of these compounds' most intriguing aspects (Ebbing/Gammon, Chapter 23).

Purpose

Using molecular models, you will study the molecular geometries of alkanes and alkenes. You will also study compounds that are derived from these by replacement of one or more hydrogen atoms with a corresponding number of hydroxyl groups. You will look for constitutional and geometric isomers for each general formula that you are given. You will also learn how to sketch a three-dimensional model in two-dimensions.

Concept of the Experiment

Many kinds of kits for making molecular models are available, so specific instructions will not be given here. Your laboratory instructor will provide instructions for using your kit. However, any molecular model that is correctly assembled will enable you to see a molecule's three-dimensional structure. Correct assembly always means that you have chosen the correct hybridization for the central atom or atoms. For example, tetrahedral sp_3 bonding cannot be used if trigonal sp^2 bonding is required.

Your molecular models will provide an easy way to look for *constitutional* and *geometric* isomers (Ebbing/Gammon, Sections 23.2 and 23.4). If you did the experiment "Geometric Isomers," you will already understand geometric isomers. Constitutional isomers are compounds with the same molecular formula but different connectivity. Make sure you are familiar with each of these types of isomers before you begin looking for them with your models.

You will also be required to provide understandable sketches. You will not need artistic or drafting ability to make these sketches. As a guide, consider methane (CH_4) and butane (C_4H_{10}) in Figure 23.1. Note that a solid line represents a bond in the plane of the paper, a wedge represents a bond coming out of the plane of the paper, and a dashed line represents a bond behind the plane of the paper. You will also note that the carbon–carbon bonds in butane have been rotated until a maximum separation between all of the hydrogen atoms has been attained. This is the most stable arrangement.

FIGURE 23.1

Sketches of methane (left) and butane (right) that are intended to show the three-dimensional aspects of the molecules.

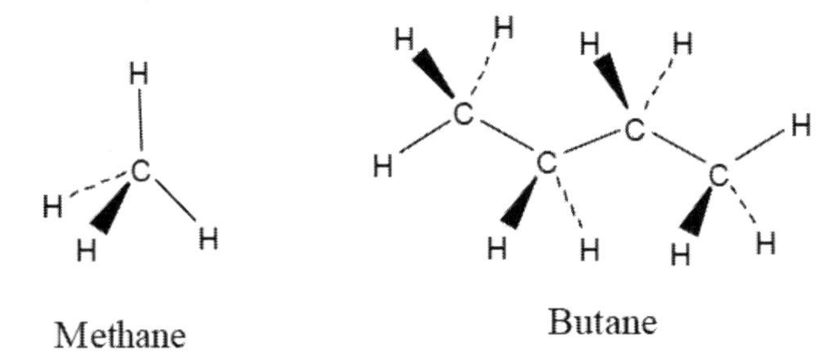

Procedure

Getting Started

1. Work with a partner.
2. Obtain a molecular model kit and instructions for its use. If possible, you and your partner should be able to examine simultaneously two models with the same general formula. This capability will help considerably when you are searching for isomers.

Doing the Experiment

1. Build molecular models representing compounds with as many of the following general formulas as you can examine.
   a. CH_4
   b. CH_3OH
   c. C_2H_6
   d. C_2H_5OH
   e. C_3H_8
   f. C_3H_7OH
   g. C_4H_{10}
   h. C_2H_4
   i. C_4H_8 (an alkene)
   j. C_2H_2 (an alkyne)

   Your laboratory instructor may give you formulas of additional compounds to model.
2. For each general formula, look for a structure, and then look for constitutional and geometric isomers. Record the results of your search.
3. Provide a structural formula for each different molecule that you discover.
4. Sketch *one* molecule from each general formula in such a way that its three-dimensional structure is apparent. Use Figure 23.1 as a guide where possible.

Date ____________ Student Name ____________
Course/Section ____________ Team Members ____________
Instructor ____________ ____________

Molecular Models of Organic Molecules

Prelaboratory Assignment

1. Provide definitions for the following terms:

   a. Structural formula

   b. Constitutional isomer

   c. Geometric isomer

2. a. Give a general molecular formula for an alkane.

   b. Give a general molecular formula for an alkene.

Date ______________________ Student Name ______________________
Course/Section ______________________ Team Members ______________________
Instructor ______________________ ______________________

Molecular Models of Organic Molecules

Results

Complete the following requirements *for each general formula.*

1. Give one structural formula.
2. Give structural formulas for all constitutional and geometric isomers that exist. If none exist, be sure to record this information.
3. Sketch one molecule. For example, if you find three constitutional isomers, you need sketch only one of these.
4. You may need to use some of your own paper to complete your study.
5. Repeat Steps 1 through 3 for any compounds that your laboratory instructor has given you.

Questions

1. Give a name to each alkane that you have found, using the IUPAC rules (Ebbing/Gammon, Section 23.5).

Student name: _______________ Course/Section: ________________ Date: ______________

2. Draw structural formulas for the following compounds:

   a. Pentane

   b. 3-methylpentane

   c. 2, 3-dimethylpentane

24A. Amylase: One of Your Enzymes

Introduction

Starch, a biological polymer (Ebbing/Gammon, Chapter 24), is one of the most important sources of carbohydrates in our diet. The digestion of starch begins in the mouth with a process called *hydrolysis* (reaction with water). This reaction, which really consists of a series of steps, is outlined in Figure 24A.1. Sugars are the ultimate products of the reaction.

The hydrolysis of starch is very slow in the absence of *amylase*, an *enzyme* found in our saliva. Enzymes are proteins that are efficient and specific catalysts of biochemical reactions. Amylase functions most efficiently at the pH and temperature of the mouth. Extreme deviation from these conditions can cause structural changes that inactivate the enzyme.

Purpose

This experiment will give you an opportunity to examine the influence of active and inactive forms of amylase on the hydrolysis of starch. Your own saliva will be the source of the enzyme.

FIGURE 24A.1

The hydrolysis of starch.

A Special Note

The activity of salivary amylase can vary widely among different individuals and for the same individual at different times. A small percentage of individuals (perhaps 3 to 4%) may find that they are unable to do this experiment at all because they have an inactive enzyme. Working with a partner who has active amylase can be an acceptable alternative.

Concept of the Experiment

Throughout this experiment, iodine will be used as an indicator for the presence of starch. Iodine and starch interact to form a blue color whose intensity is proportional to the quantity of starch that is present. During hydrolysis, the quantity of starch will decrease as the reaction progresses, and the blue color will fade. If the intensity of the blue color does not diminish, hydrolysis has not occurred (at least not to any appreciable extent).

You will use four dilute saliva solutions, called A, B, C, and D, in this experiment. Their uses are outlined in Table 24A.1. You will use one or more numbered test tubes with each of these solutions. These numbers are also given in the table. You can avoid confusion about the ultimate purpose of any solution by referring to this table.

Table 24A.1

Purpose of Solutions

Solution	Use	Test Tube No.
A	Qualitative rate study	1–6
B	Effect of acid	7
C, D	Effect of heat	8

Solution A will be used to examine the progress of the enzyme-catalyzed hydrolysis of starch during a period of 10 min. The activity of your salivary amylase will become apparent during this part of the experiment.

Solution B will be used to determine the effect of acid on the active enzyme, and solutions C and D will be used to determine the effect of heat. If acid or heat makes your amylase totally inactive, catalysis will no longer occur.

Procedure

Getting Started

1. Obtain about 2 mL of your saliva in a 5-mL or 10-mL graduated cylinder.
2. Transfer the saliva to a clean Erlenmeyer flask containing 12 mL of distilled water. Agitate the flask thoroughly to homogenize the solution.
3. Mark 2 clean beakers with identifying letters (A and B).
4. Obtain 7 small test tubes.
5. Mark 6 of these test tubes with identifying numbers (1 through 6). Add 1 mL of distilled water to each, and mark the height of the water with a marking pencil. These marks will enable you to add 1 mL of a solution quickly without using a graduated cylinder. Pour the water into a sink.
6. Mark the remaining test tube with an identifying letter (C).
7. Set up a ring stand, an iron ring, a wire gauze, and your laboratory burner. Adjust the height of the ring so that the wire gauze will be in the hottest part of the flame from the burner. Do not light the burner until you have made this adjustment.

   CAUTION: Do not touch the iron ring at any time after the burner has been lit. Avoid burning your fingers. Remember, hot glass looks just like cold glass. Report any burns, and treat them by immersing in ice water.

8. Place a 250-mL beaker containing distilled water on the wire gauze. Heat the water to a gentle boil.
9. Handle the solution of hydrochloric acid carefully.

   CAUTION: A 6 *M* solution of hydrochloric acid can cause chemical burns, in addition to ruining your clothes. Be sure to wear approved chemical splash goggles. If you spill any of this solution on you, wash the contaminated area thoroughly and immediately report the incident to your laboratory instructor. You may require further treatment. You might wish to wear gloves if they are available in the laboratory.

Preparing Solution C

1. Add 1 mL of the saliva solution to test tube C. This is solution C.
2. Heat it in the boiling water for 20 min. During this time, go on to the next part of this experiment.

   However, when 20 min have elapsed, remove the test tube from the bath and allow the saliva solution to cool.

Working with Solution A

1. Place 6 mL of a 1% starch solution into beaker A. This is solution A.
2. Add 1 drop of the iodine solution and 1 drop of 6 *M* HCl to each of test tubes 1 through 6.
3. Pour 1 mL of solution A into test tube 1, using the mark that you have made to measure the volume. *Save this test tube and its contents.* You will use it as a reference throughout the experiment.
4. Read Steps 5 through 9 before proceeding.
5. Add 1 mL of the saliva solution to solution A, using a 5-mL or 10-mL graduated cylinder and noting the time to the nearest second. Swirl the solution immediately and thoroughly.
6. When 1 min has elapsed, pour 1 mL of the solution from Step 5 into test tube 2. Shake the test tube gently.
7. Repeat Step 6 after 2, 4, 6, and 10 min, using test tubes 3, 4, 5, and 6, respectively.
8. Compare the colors in test tubes 1 through 6, and record the results. If the color has lessened as time has progressed, hydrolysis of starch has occurred. Record the degree of hydrolysis (none, some, or complete).
9. *Save test tube 6 with its solution.* You will use it as a reference throughout the remainder of this experiment. This test tube shows the maximum amount of hydrolysis that can occur during 10 min with your amylase.
10. Discard the solutions in test tubes 2 through 5 as directed by your laboratory instructor. Wash, rinse, and dry test tubes 2 through 5. Renumber two of them (7 and 8).

Working with Solution B

1. Obtain 6 mL of the 1% starch solution in beaker B. This is solution B.
2. Add 4 drops of 6 *M* HCl to solution B.
3. Add 1 mL of the saliva solution to solution B. Note the time while swirling to obtain a homogeneous solution.
4. Set the solution from Step 3 aside for exactly 10 min.
5. Add 1 drop of iodine solution to test tube 7.
6. After 10 min have elapsed, pour 1 mL of the solution from Step 4 into test tube 7. Shake the test tube gently.
7. Record the color in test tube 7. Compare the color with those in test tubes 1 and 6. Has any hydrolysis occurred?
8. Discard any remaining solution in beakers A and B. Wash, rinse, and dry beakers A and B. Identify one of them with a new letter (D).

Working with Solutions C and D

1. Obtain 6 mL of the 1% starch solution in beaker D. This is solution D.
2. When solution C has cooled to room temperature, add it to solution D, noting the time. Swirl.
3. Set solution D aside for exactly 10 min.
4. Add 1 drop of iodine solution to test tube 8.
5. Pour 1 mL of the solution from Step 2 into test tube 8. Shake the test tube gently.
6. Record the color in test tube 8. Compare the color with those in test tubes 1 and 6. Has any hydrolysis occurred?

CAUTION: Before you leave the laboratory, make sure that your gas outlet and those of your neighbors are closed.

Date ______________________ Student Name ______________________
Course/Section ______________________ Team Members ______________________
Instructor ______________________ ______________________

Amylase: One of Your Enzymes

Prelaboratory Assignment

1. Provide definitions for the following terms:

   a. Enzyme

   b. Catalyst

   c. Amylase

   d. Hydrolysis

2. How will you know if the hydrolysis of starch has occurred in this experiment?

3. Provide a detailed description of the purposes of the contents of the 8 test tubes listed in Table 24A.1.

4. What special safety precautions are cited in this experiment?

Date ____________________ Student Name ____________________
Course/Section ____________________ Team Members ____________________
Instructor ____________________ ____________________

Amylase: One of Your Enzymes

Results

1. *Solution A*

Test Tube	Reaction Time (min)	Description of Color	Degree of Hydrolysis
1	0	________	
2	1	________	________
3	2	________	________
4	4	________	________
5	6	________	________
6	10	________	________

2. *Solution B*

Test Tube	Reaction Time (min)	Description of Color	Degree of Hydrolysis
7	10	________	________

3. *Solutions C and D*

Test Tube	Reaction Time (min)	Description of Color	Degree of Hydrolysis
8	10	________	________

Questions

1. a. Summarize the effects of acidity and heat on your amylase.

   b. What reason can you give for the effects of acidity and heat on the activity of your amylase?

24B. Catalase and the Decomposition of Hydrogen Peroxide

Introduction

When hydrogen peroxide decomposes, the products are oxygen and water. Although this reaction is spontaneous in the sense of thermodynamics, it is very slow under ordinary conditions. The rate is accelerated, however, by such catalysts as aqueous Fe^{3+} and I^- ions. An enzyme called *catalase* is a biological catalyst for this reaction. This protein (Ebbing/Gammon, Chapter 24) is such an effective catalyst that 1 mg will cause the same rate of decomposition of hydrogen peroxide as 2 kg of Fe^{3+} ions.

Catalase, a protein with a molecular weight of about 5.4×10^5 g/mol, contains a heme group like the one found in hemoglobin. It occurs in such diverse things as mammalian liver, potatoes, and baker's yeast. Baker's yeast will be used in this experiment.

Purpose

During this experiment, you will measure the amount of oxygen that is released during the catalyzed decomposition of hydrogen peroxide. This measurement will allow you to calculate the original concentration of the hydrogen peroxide.

Concept of the Experiment

The quantity of oxygen that is evolved will enable you to determine the molarity of the solution of hydrogen peroxide (H_2O_2). To do so, you will need to calculate the number of moles of gas by using the volume of the evolved gas, the atmospheric pressure, Dalton's law of partial pressures (you will be

When you use Dalton's law of partial pressures, you will need the vapor pressure of water. The vapor pressure of water at various temperatures can be found in Table 5B.1 in the experiment "The Decomposition of Potassium Chlorate."

Procedure

Getting Started

1. Work with a partner.
2. Obtain a 10-mL transfer pipet, a thermometer, a vial (fitted with a short length of copper wire), a 250-mL Erlenmeyer flask, a 500-mL Erlenmeyer flask, a 400-mL beaker, two stoppers with attached plastic tubing, and a screw pinch clamp.
3. Obtain about 25 mL of the H_2O_2 solution whose concentration is to be determined. You also will need a small quantity of baker's yeast.
4. Obtain and record the atmospheric pressure in the laboratory.

Assembling the Apparatus

1. Use the diagram in Figure 24B.1 as a guide in assembling the apparatus. Use the 250-mL Erlenmeyer flask as flask A, the 500-mL Erlenmeyer flask as flask B, and the 400-mL beaker as beaker D. Put a pinch clamp near the beaker end of tube C.

2. Clamp flask A to a ring stand.
3. Add 400 mL of water to flask B and 300 mL of water to beaker D. Mark the water level in flask B with a marking pencil.
4. Place the entire apparatus at least 1 ft from the edge of the laboratory bench to minimize the chance of knocking the apparatus off the bench.
5. Fill tube C with water. To do so, remove the stopper from flask A and loosen the pinch clamp on tube C. Draw water into the tube from the beaker by using a rubber suction bulb on the glass tube that was in flask A. If air bubbles remain in the tube, they may be removed by siphoning water back and forth from flask B to beaker D. Siphoning will occur when the flask and the beaker are alternately raised and lowered. *Do not try to siphon by sucking on the tube with your mouth.*

FIGURE 24B.1

The arrangement of glassware and tubing.

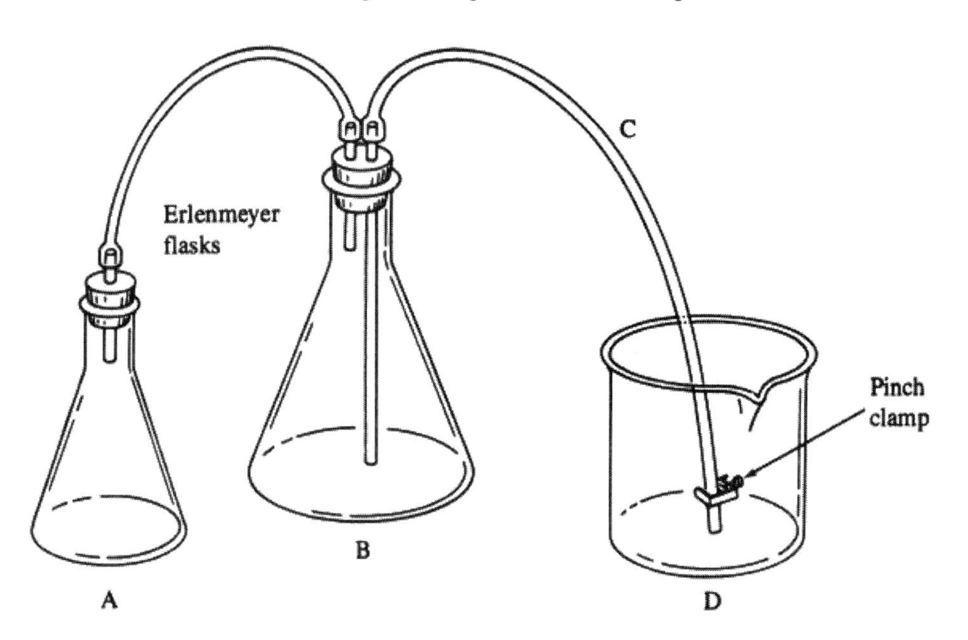

6. Use siphoning to bring the water level in flask B back to the 400-mL mark that you placed on the flask. Tighten the pinch clamp while the water is at that level.

Testing for Leaks

1. Return the stopper to flask A, putting it *firmly* in place, and open the pinch clamp.
2. Raise beaker D. If water flows continuously from the beaker to flask B, there is a leak in your apparatus. Make sure the stoppers in the flasks are pushed in as much as possible. If necessary, ask your laboratory instructor for help.
3. When your apparatus is free of leaks, bring the water level in flask B back to the 400-mL mark that you made.
4. Tighten the pinch clamp.

Setting Up the Vial

1. Angle the copper wire attached to the vial so that it is about 10-20° from a vertical position, as shown in Figure 24B.2. The wire will prevent the vial from tipping over when the vial is in flask A. However, the bent wire will allow the vial to tip slightly so that a solution in flask A can enter the vial when the flask is tilted.

2. Remove the stopper from flask A and add 135 mL of water.

3. To test the position of the wire, use crucible tongs to lower the vial gently into flask A until the *upright* vial rests on the bottom of the flask. Use the tongs to tip the vial until the wire touches the wall of the flask. The mouth of the vial should be above the water level in the upright flask. When the flask is tilted, water should flow into the vial if the wire has been bent properly.

FIGURE 24B.2

The vial and the attached piece of copper wire.

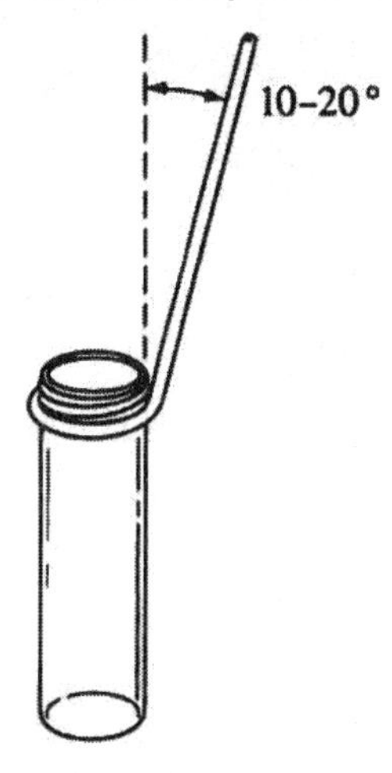

4. Once the wire is positioned properly, remove the vial from the flask, discard the water in the vial and the flask, and dry the interior and exterior of the vial.

Doing the Reaction

1. Pipet 10.0 mL of the solution of H_2O_2 into flask A.

2. Add 125 mL of distilled water to flask A from a graduated cylinder. Swirl gently to obtain a homogeneous solution.

3. Using a metal spatula, add baker's yeast to the vial to a depth of about 1/4 inch. Wipe the exterior of the vial carefully to remove any of the yeast that may be there.

4. Make sure that flask A is clamped to the ring stand.

5. Using crucible tongs, lower the vial gently into flask A until the *upright* vial rests on the bottom of the flask.

6. Return the stopper to flask A, and recheck the system for leaks with the method that you used earlier.

7. Adjust the pressure inside the apparatus to atmospheric pressure with the following procedure.

   Open the pinch clamp. Raise beaker D until the water level in that beaker is exactly equal to that in flask B. Water will siphon into or out of the beaker until the internal and external pressures are equal. Close the clamp tightly.

8. Pour all of the water out of beaker D, but do *not* dry the beaker. Place the end of tube C (still filled

with water) back in the beaker.

9. Keeping the pinch clamp closed for the moment, remove flask A from its clamp, and tilt the flask carefully so that the solution in the flask flows into the vial. Do not allow the solution to touch the rubber stopper.
10. Gas evolution will begin immediately. After about 1 min, loosen the pinch clamp. Water will be forced from flask B to beaker D. Agitate flask A gently and continuously until no more effervescence (bubbling) occurs.
11. Clamp flask A to the ring stand and allow another 5 min for the heat generated by the reaction, if any, to dissipate.
12. Adjust the pressure inside the apparatus to atmospheric pressure, using the method in Step 7.
13. Loosen the stopper in flask B, insert the thermometer into the flask, and measure the temperature of the gas (not the water). The temperature of the gas should agree with the temperature of the water in beaker D to within 2°C. If it does not, more time should have been allowed for the heat of the reaction to dissipate. Record the temperature of the gas to the nearest degree.
14. Carefully measure the volume of the water in the beaker using a 1-L graduated cylinder. Record the result.
15. Discard the solution in flask A, rinse the flask and the vial, dry the interior and exterior of the vial, and replace the water in flask B and beaker D.
16. Repeat Steps 1 through 14. If the volumes of the evolved gas in the two trials differ by more than 10 mL, perform another trial.

Date ______ Student Name ______
Course/Section ______ Team Members ______
Instructor ______ ______

Catalase and the Decomposition of Hydrogen Peroxide

Prelaboratory Assignment

1. a. What chemical reaction will occur during this experiment?

   b. What quantities must be measured in this experiment?

2. a. Why is it necessary to equalize the water levels in flask B and beaker D before you measure the volume of displaced water?

   b. Just before you allow the reaction to begin, you are asked to discard the water in beaker D. Why are you forbidden to dry the beaker?

3. A quantity of O_2 is collected over water at 23°C and an atmospheric pressure of 744 mmHg. What is the pressure of the gas?

Date	______________	Student Name	______________
Course/Section	______________	Team Members	______________
Instructor	______________		______________

Catalase and the Decomposition of Hydrogen Peroxide

Results

Atmospheric pressure (mmHg): ______________

Trial	1	2	3
Volume of H_2O_2 solution (mL)	________	________	________
Temperature of the gas (°C)	________	________	________
Volume of water displaced (mL)	________	________	________
Vapor pressure of water (mmHg)	________	________	________
Pressure of the gas (mmHg)	________	________	________

Questions

1. a. Calculate the number of moles of gas that evolved in each trial.

b. Using the mean number of moles of the gas, calculate the molarity of the H_2O_2 solution.

2. The solutions of hydrogen peroxide that we can purchase in a pharmacy or a supermarket are 3% H_2O_2 by mass. Using any reasonable assumption that you wish, calculate the percentage of H_2O_2 by mass in the solution that you used in this experiment. Make sure that you state your assumption.

Date ______________________ Student Name ______________________
Course/Section ______________________ Team Members ______________________
Instructor ______________________ ______________________

Inquiries with Limited Guidance

Inquiry 1

Demonstration

In a hood, pipet 5 mL of acetone into a 10-mL graduated cylinder, followed by 5 mL of water from a pipet. Note the effect.

CAUTION: ACETONE IS FLAMMABLE. AVOID FLAMES

Inquiry

Is it true that the volume of a solution of two liquids is always the sum of the volumes of the unmixed liquids? Investigate the additivity of the volumes.

Date ______________________ Student Name ______________________

Course/Section ______________________ Team Members ______________________

Instructor ______________________ ______________________

Inquiries with Limited Guidance

Inquiry 2

Demonstration

Using tongs, ignite a small piece of magnesium with a burner. Note the effect.

CAUTION: AVOID BURNING YOUR FINGERS. AVOID LOOKING DIRECTLY AT THE FLAME.

Inquiry

Investigate the enthalpy of combustion of the metal.

Date ______________________ Student Name ______________________
Course/Section ______________________ Team Members ______________________
Instructor ______________________ ______________________

Inquiries with Limited Guidance

Inquiry 3

Demonstration

Fill the chamber of an alcohol burner with ethanol, and ignite the wick.

CAUTION: AVOID BURNS. Alcohols and other volatile solvents have vapor trails, which can ignite at a flame that is some distance from the solvent and flash back to the solvent resulting in a larger fire and possible explosion. Be especially careful if removing alcohol from the burner.

Inquiry

Both methanol and ethanol have been discussed as possible fuels for automobile engines. Investigate which of these alcohols will provide more heat energy on a mass basis, a volume basis, and a mole basis.

Date ____________________ Student Name ____________________
Course/Section ____________________ Team Members ____________________
Instructor ____________________ ____________________

Inquiries with Limited Guidance

Inquiry 4

Demonstration

Drop part of a Rolaids® tablet into a small amount of (1 *M*) hydrochloric acid. Note the effect.

Inquiry

The manufacturers have claimed that one of its tablets will consume 47 times its weight in stomach acid? Investigate this claim.

Date	______________________	Student Name	______________________
Course/Section	______________________	Team Members	______________________
Instructor	______________________		______________________

Inquiries with Limited Guidance

Inquiry 5

Demonstration

Add two drops of phenolphthalein to 2 mL of 0.1 *M* HCl, and then add drops of 0.1 *M* NaOH. Repeat, substituting 0.1 *M* acetic acid for HCl. Note the effects. Repeat, using a mixture of HCl and acetic acid instead of a single acid.

Inquiry

Investigate the claim that in a mixture of these two acids, the concentration of each of the acids can be determined.

Date ____________________ Student Name ____________________
Course/Section ____________________ Team Members ____________________
Instructor ____________________ ____________________

Inquiries with Limited Guidance

Inquiry 6

Demonstration

Add 2 drops of methyl red indicator to 2 mL of 0.1 *M* acetic acid, and then add drops of 0.1 *M* NaOH until you see an effect. Then add drops of 0.1 *M* HCl.

Inquiry

Investigate the applicability of the Henderson–Hasselbalch equation to the titration of acetic acid with sodium hydroxide.

Date		Student Name	
Course/Section		Team Members	
Instructor			

Inquiries with Limited Guidance

Inquiry 7

Demonstration

Add 2 drops of bromocresol green to 1 mL of 0.1 M HCl, and then add drops of 0.1 M NaOH until you see an effect.

Inquiry

Investigate using the color of a solution of bromocresol green to determine the pH.

Date ____________________ Student Name ____________________
Course/Section ____________________ Team Members ____________________
Instructor ____________________ ____________________

Inquiries with Limited Guidance

Inquiry 8

Demonstration

Add a few drops of a solution of sodium sulfate to a few drops of a solution of barium nitrate.

Inquiry

Investigate the identity of the precipitate qualitatively and quantitatively.

Date ____________________ Student Name ____________________

Course/Section ____________________ Team Members ____________________

Instructor ____________________ ____________________

Inquiries with Limited Guidance

Inquiry 9

Demonstration

Measure the voltage of a voltaic cell composed of Cu^{2+} (0.1 *M*) | Cu and Zn^{2+} (0.1 *M*) | Zn electrodes.

Inquiry

Investigate the effect of adding aqueous ammonia (with stirring) to the Cu^{2+} (0.1 *M*) | Cu electrode and the cause of any observed effect..

Date ______ Student Name ______
Course/Section ______ Team Members ______
Instructor ______ ______

Inquiries with Limited Guidance

Inquiry 10

Demonstration

Add drops of the solution of the red food dye to 2 mL of the solution of the blue food dye. Note the effect.

Inquiry

Investigate determining the concentrations of the two dyes in a mixture by using the color of the mixture.

Appendixes

Appendix A: Mistakes, Errors, Accuracy, and Precision

Mistake and *error* have entirely different meanings in the language of science. A mistake is a blunder or unintentional action whose consequence is undesirable. Error, on the other hand, accounts for the range of values obtained from successive measurements of the same quantity, even though there was no mistake in any of the measurements. Moreover, error may be either *systematic* or *random.*

Systematic Error

A systematic error causes a measurement to be always too large or, alternatively, always too small. An error of this type can be caused by a faulty measuring device. For example, if the intervals between the millimeter markings on a ruler are always 9/10 of the correct interval, any measurement made with that ruler will always be too small and inaccurate. Systematic error can also be caused by the consistent incorrect use of a very good measuring device.

Systematic errors influence the *accuracy* of a measurement, or, in other words, the agreement between a measured value of a quantity and its true value. In cases in which it is not possible to know the true value of a quantity, it is impossible to determine the accuracy of the measurement.

Random Error

Random errors are observed when a measuring device, even a very accurate one, is used a number of times to make the same measurement. If a very large number of these measurements were made, the results could be described by the bell-shaped curve shown in Figure A.1. This graph is obtained by plotting an individual measurement value against the number of times that this value was observed. The most frequently observed values are those around the midpoint, whereas rarely occurring values of the measurement are found at either end of the curve. The average value, or mean value, of this set of measurements corresponds to the maximum height of the curve.

The random errors that cause curves such as the one shown in Figure A.1 are closely linked to the *precision* of the measurements. When most of the measurements have values closely dispersed around the mean because of high precision, a very narrow and steep curve results. Low precision causes a widely dispersed curve with a low maximum, because the values of many of the measurements may differ significantly from the mean. A comparison of the curves that result from precise and imprecise measurements is shown in Figure A.2.

FIGURE A.1

A graph showing the distribution of experimental results when a single quantity is measured many times. The value of a measurement is plotted against the number of times that this value was found. The average or mean value of the measurement occurs at the maximum.

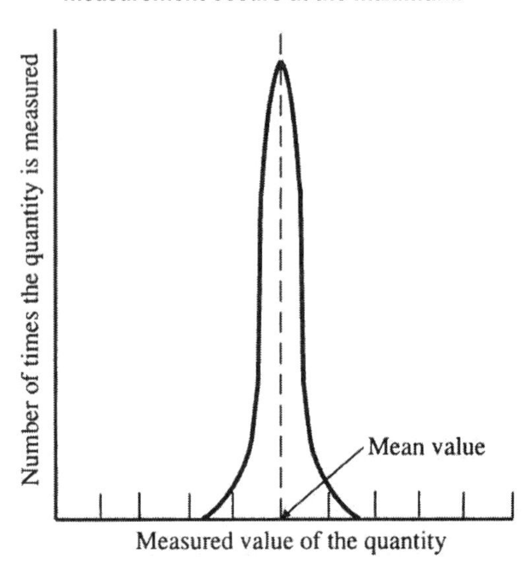

FIGURE A.2

A comparison of two distributions of experimental results. Both have the same mean value, but the measurements have high precision in curve A and lower precision in curve B.

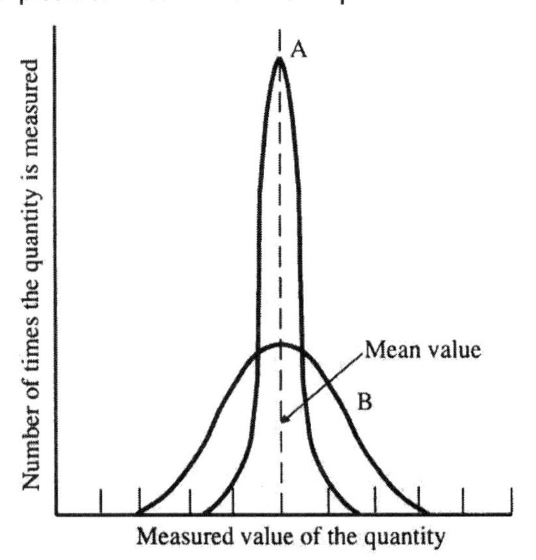

A measuring device does not need to be accurate to permit the high precision found in Figure A.2A. Because there is no necessary connection between accuracy and precision, it is possible to obtain measurements of high precision with a faulty measuring device.

Precision and Standard Deviation

Precision is the dispersion of, or closeness of the agreement between, successive measurements of the same quantity. The dispersion in a set of measurements is usually expressed in terms of the *standard deviation,* whose symbol is s:

$$s = \left(\frac{\Sigma d_i^2}{N - 1}\right)^{1/2}$$

where

Σ means "the sum of"

$d_i = x_i - \bar{x}$ = deviation

x_i = a particular value of a measurement

$\bar{x}$ = the mean value

N = the number of measurements

Measurements with high precision (Figure A.2A) are narrowly dispersed, and these measurements have a smaller standard deviation than measurements with lower precision (Figure A.2B).

Unless the number of measurements of the same quantity is very large, the calculated value of the standard deviation is only an estimate of the true standard deviation. Nevertheless, even a limited set of measurements will allow an estimate of the dispersion in the measurements to be judged.

How is the formula used? The formula states: Find the sum of the squares of the deviations, divide by one less than the total number of measurements, and take the square root of the result. The following example illustrates the use of the formula. Suppose we measure the length of an object seven times with a ruler. The values of these measurements (x_i) are 10.11 cm, 10.13 cm, 10.10 cm, 10.12 cm, 10.15 cm, 10.11 cm, and 10.12 cm. The calculation of the standard deviation of these results is shown in Table A.1.

Table A.1

An Example of Calculating a Standard Deviation

Value of Measurement x_i	Deviation $d_i = (x_i - \bar{x})$	Squared Deviation d_i^2
10.11	−0.01	0.0001
10.13	+0.01	0.0001
10.10	−0.02	0.0004
10.12	0.00	0.0000
10.15	+0.03	0.0009
10.11	−0.01	0.0001
10.12	0.00	0.0000
Sum = 70.84		Sum = 0.0016

Mean $(\bar{x}) = \Sigma x_i = \frac{70.84}{7} = 10.12$

$s = \left(\frac{0.0016}{7 - 1}\right)^{1/2} = 0.016$

When a quantity, such as the length of an object, is measured several times, it is customary to report the mean value of the measurements. The dispersion, or precision, of the measurements can be indicated, according to one custom, by writing $\pm$ the calculated value of s after the mean. Thus we would report 10.12 ± 0.02 cm for the example in Table A.1. Note that the standard deviation of 0.016 was rounded to two figures because there are only two figures to the right of the decimal point in the mean. When we report 10.12 ± 0.02 cm as the best value for the quantity, we are stating that the length of the object probably lies between

$$10.12 + 0.02 = 10.14 \text{ cm}$$

and

$$10.12 - 0.02 = 10.10 \text{ cm}$$

Precision and Significant Figures

The precision of a set of measurements can also be gauged by the number of significant (that is, meaningful) figures that are used in the mean value of the measurements. This is the method that is required in the experiment "Some Measurements of Mass and Volume."

The correct number of significant figures in a mean value will always be the number of certain digits plus one uncertain digit. In the example in Table A.1, we have shown that the length of the object probably lies between 10.10 cm and 10.14 cm, with a mean value of 10.12 cm. Clearly, the first three digits in the mean value are certain, and uncertainty occurs in the fourth digit. The precision of these measurements justifies the use of four significant figures in the mean value.

It will be instructive to consider one further example. Suppose we measure the mass of an object six times and find values of 13.34 g, 13.08 g, 13.58 g, 13.42 g, 13.29 g, and 13.45 g. As you should verify by calculations, the mean value is 13.36 g and the standard deviation is 0.17 g. The mass of the object probably lies somewhere between 13.19 g and 13.53 g. In this case, the first uncertain digit is the third digit. The precision of these measurements, therefore, will justify only three significant figures, even though four digits were obtained in each measurement.

An important point should be noted here. The precision that we obtain in making a measurement is a characteristic property of the measuring device. For example, consider the ruler that was used to obtain the data in Table A.1. This device will always give a measurement in which the digit in the place immediately after the decimal point is certain, no matter what object is being measured. Moreover, the first uncertain digit will always occur in the hundredths place. As a result, the mathematical procedure for determining the precision of a measurement is required only the first time you use a measuring device. After the first time, you should know the precision that you can obtain with that device.

Counting Significant Figures

We will observe the following rules when we need to count the number of significant figures in a measured quantity (Ebbing/Gammon, Section 1.5):

1. All digits are significant except zeros at the beginning of a number and possibly terminal zeros. Thus 5.46 cm, 0.546 cm, and 0.00546 cm all contain three significant figures.
2. Terminal zeros that occur to the right of a decimal point are taken to be significant. Each of the following has four significant figures: 14.10 cm, 141.0 cm, and 14.00 cm.
3. Internal zeros are significant. Thus the zero in 10.3 is significant, and there are three significant figures.
4. Terminal zeros to the left of the decimal point are ambiguous (i.e. may or may not be significant) and should be avoided. Ambiguity can be removed by using scientific notation.

Significant Figures in Calculations

Measured quantities are often used in calculations. After we complete the calculation, how many significant figures should appear in the answer? Suppose we want to determine the area of a rectangle. We measure the length and width and find these dimensions to be 115.36 cm and 3.52 cm, respectively. The area is equal to the length multiplied by the width. When we do this calculation with a pocket calculator, we obtain 406.0672 cm2. It is incorrect, however, to use this number. The reason is simple: Its implied precision (seven significant figures) is much greater than the precision of the numbers used to obtain it.

In general, the precision of an answer to a calculation cannot exceed the precision of the measured quantities used in the calculation. We will apply two different and distinct rules to achieve this result (Ebbing/Gammon, Section 1.5):

1. When measured quantities are *multiplied* or *divided*, there should be as many significant figures in the answer as there are in the measurement with the least number of significant figures. In the calculation of the area that we just did, 3.52 cm has the least number of significant figures (three). Therefore, the answer should be reported to three significant figures, or 406 cm^2.
2. When measured quantities are *added* or *subtracted*, a different rule applies. There should be the same number of decimal places in the answer as there are in the measurement with the least number of decimal places. Suppose we wish to add 103.1 cm and 0.334 cm. Strictly speaking, the result would be 103.434 cm. But because the quantity 103.1 cm has only one decimal place whereas 0.334 cm has three, the answer is 103.4 cm.

It is important to note that any number whose value is known exactly will not affect the number of significant figures in a calculated result. For example, there are *exactly* 1×10^2 centimeters in 1 meter, so

$$1 \times 10^2 \text{ cm/m} \times 5.243 \text{ m} = 5.243 \times 10^2 \text{ cm}$$

The number of significant figures in the answer is determined by 5.243, not 1×10^2.

Rounding

Rounding is often required to obtain the correct number of significant figures. We will use the following general procedure (Ebbing/Gammon, Section 1.5). Look at the leftmost digit to be dropped.

1. If this digit is 5 or greater than 5, add 1 to the last digit to be retained, and drop all digits further to the right. Thus rounding 1.2151 to three significant figures gives 1.22.
2. If this digit is less than 5, drop it and all digits further to the right. Rounding 1.2143 to three significant figures gives 1.21.

In doing a calculation of two or more steps, it is desirable to retain additional digits for intermediate answers. This ensures that small errors from rounding do not accumulate in the final result. If you use a calculator, you can simply enter numbers one after the other, performing each arithmetic operation and rounding just the final answer.

Appendix B: Using a Coffee-Cup Calorimeter

Polystyrene coffee cups make excellent calorimeters because of their ability to block the passage of heat. The complete calorimeter consists of two nested 6-ounce cups, a top, a thermometer, and a stirring device. The example shown in Figure B.1 is stirred by a plastic-encased iron bar that, in turn, is moved by a magnetic stirrer. However, stirring by hand yields the same results.

FIGURE B.1

The coffee-cup calorimeter. Mechanical stirring also can be used.

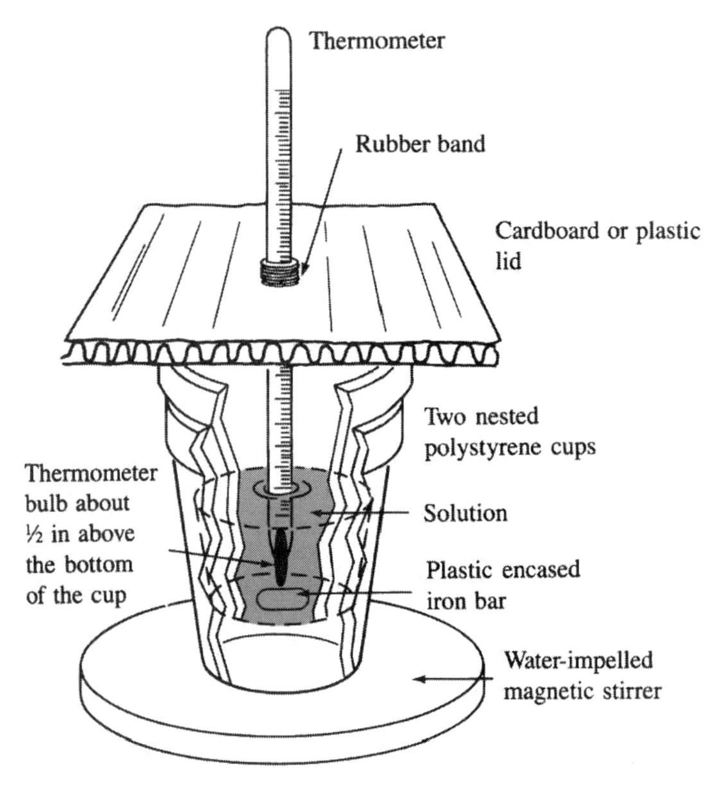

Like any other calorimeter, the coffee-cup calorimeter provides the means to measure the heat flow between a system and its surroundings. The meaning of these words must be understood in terms of our calorimeter.

Defining the Surroundings and the System

We must begin by defining how much of the surroundings we will be required to consider. Can we limit the surroundings to a small region, or must we consider the entire laboratory?

The problem becomes rather simple if we assume that our calorimeter is perfectly insulated. Heat, we assume, will not flow through the walls of the calorimeter. This assumption allows us to restrict the extent of the surroundings. Because heat cannot flow out of or into the calorimeter, we can define the surroundings as the complete calorimeter and any water *whose temperature is initially identical to that of the calorimeter.* Later, we will find a way to correct for imperfect insulation without altering our definition of the surroundings.

The system includes any other substance or substances that are contained in the calorimeter. This definition of a system includes substances that are dissolved in the water, such as the reactants and products of a chemical reaction. It also includes other portions of water whose temperature is not initially identical to the temperature of the calorimeter.

An Equation for Heat Flow

The equation that describes heat flowing between a system and its surroundings is

$$q(\text{system}) = -q(\text{surroundings})$$

This equation states that the heat lost (or gained) by a system, q(system), is equal to the heat gained (or lost) by the surroundings, $-q$(surroundings). Clearly, q(system) and q(surroundings) must have opposite signs, because as heat is lost by one, it is gained by the other.

Because of our definition of the surroundings, the equation for heat flow becomes

$$q(\text{system}) = -q(\text{water}) - q(\text{calorimeter})$$

The following equations show how the heat gained or lost by the water and the calorimeter, q(water) and q(calorimeter), can be calculated:

$$q(\text{water}) = \text{sp. ht.} \times \text{mass of } H_2O \times (t_f - t_i)$$

$$q(\text{calorimeter}) = C \times (t_f - t_i)$$

where sp. ht. is the specific heat, C is the heat capacity of the calorimeter, t_f is the final temperature, and t_i is the initial temperature. We will take the specific heat to be that of pure water, 4.184 J/(g • °C). Unless we are dealing with pure water, however, this is only an approximation. When substances are dissolved in the water, its specific heat is altered somewhat. We will also take the heat capacity of the calorimeter to be 1.0×10^1 J/°C.

An Example

Suppose the temperature of 50 mL of 1.0 *M* NaOH in a coffee-cup calorimeter is 25.3°C. When 50 mL of 1.0 *M* HCOOH (formic acid), whose temperature is also 25.3°C, is added to the calorimeter, the temperature increases to 31.8°C. The chemical reaction is

$$NaOH + HCOOH \rightarrow NaHCOO + H_2O$$

The density of the final solution can be taken to be 1.0 g/mL. We will identify the surroundings and the system and then calculate q(system).

The surroundings are the complete calorimeter and all the water from both of the original solutions. Note that this water and the calorimeter share the same temperature. Thus the demands of our definition of the surroundings are met. The mass of this water is

$$(50 + 50) \text{ mL} \times 1.0 \text{ g/mL} = 1.0 \times 10^2 \text{ g}$$

The system becomes the reactants and products of the chemical reaction, including the quantity of water that is formed in the reaction.

We will now calculate q(water) and q(calorimeter) from our equation for heat flow. We will use 4.184 J/(g • °C) as the specific heat of water and 1.0 × 101 J/°C as the heat capacity of the coffee-cup calorimeter.

$$\begin{aligned} q(\text{water}) &= 4.184 \text{ J/(g} \cdot {}^\circ\text{C)} \times 1.0 \times 10^2 \text{ g} \times (31.8 - 25.3)^\circ\text{C} \\ &= 2.7 \times 10^3 \text{ J} \\ q(\text{calorimeter}) &= 1.0 \times 10^1 \text{ J/}^\circ\text{C} \times (31.8 - 25.3)^\circ\text{C} \\ &= 65 \text{ J} \end{aligned}$$

Then

$$\begin{aligned} q(\text{system}) &= -q(\text{water}) - q(\text{calorimeter}) \\ &= -2.7 \times 10^3 \text{J} - 65 \text{ J} \\ &= -2.8 \times 10^3 \text{J or } -2.8 \text{ kJ} \end{aligned}$$

Enthalpy Changes

Enthalpy changes can be calculated by dividing q(system) by the number of moles of the limiting reactant. In our example, we have, for each reactant,

$$50 \text{ mL} \times 1 \text{ L}/10^3 \text{ mL} \times 1.0 \text{ mol/L} = 5.0 \times 10^{-2} \text{ mol}$$

The enthalpy change becomes

$$\Delta H = -2.8 \text{ kJ}/5.0 \times 10^{-2} \text{ mol} = -56 \text{ kJ/mol}$$

Correcting for Imperfect Insulation

The equation for heat flow was obtained using the assumption that our calorimeter is perfectly insulated. We must now recognize that this assumption is not warranted, because heat will flow through the walls. After all, hot coffee in a polystyrene coffee cup cools even if the top is covered.

Because heat leaks through the walls, we will not be able to observe the highest (or lowest) temperature that could have been achieved in a perfectly insulated calorimeter. However, we need to know that temperature, because it is t_f in our equation for heat flow.

We will estimate that temperature by plotting temperature as a function of time. We will then use the data to *extrapolate* to the temperature at the time at which the solutions were mixed. This temperature is t_f. A typical result is shown in Figure B.2, which indicates how a final temperature of 31.8°C was estimated in the preceding example. Note, however, that the rate and method of stirring affect the appearance of this graph.

FIGURE B.2

A graph showing the temperature as a function of time after 50 mL of 1.0 *M* NaOH and 50 mL of 1.0 M HCOOH are mixed.

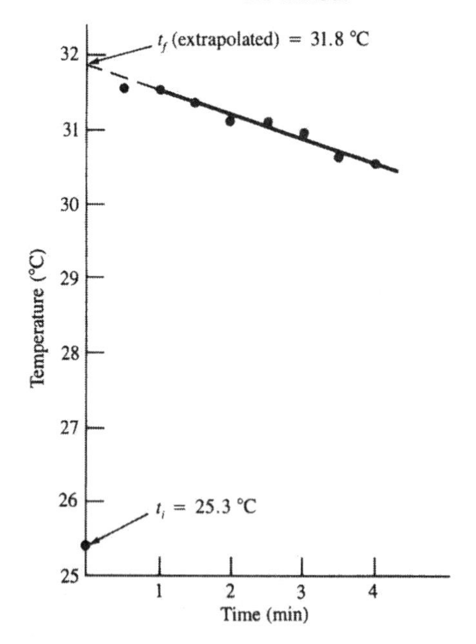

Appendix C: The Absorption of Light

When a substance absorbs light, an electron undergoes a transition from the lowest-energy level to a higher-energy level. The energy gap between these levels is given by Einstein's equation $E = h\upsilon$. Consequently, only one frequency υ causes this transition. More often than not, a chemist refers to the wavelength of light (λ) rather than to its frequency. Einstein's equation then becomes $E= hc/\lambda$, where c is the velocity of light.

Transmittance and Absorbance

When light of the correct wavelength shines through a solution of a substance that absorbs light, the intensity of the light diminishes as it passes through the solution because absorption occurs. If the intensities of the light that enter and emerge from the solution are represented by $I0$ and I, respectively, *transmittance* (T) is defined as the ratio

$$T = I/I_0$$

A related quantity called the *absorbance* (A) is defined as the negative logarithm of the transmittance.

$$A = -\log_{10} T = -\log_{10}(I/I_0)$$

Spectrophotometers

Absorbance is measured with an instrument called a spectrophotometer. This instrument separates light into its component wavelengths and selectively measures the intensity of the light of a given wavelength after it passes through a solution. All spectrophotometers, regardless of the manufacturer, have certain common fundamental parts. These parts include a source of radiant energy, a prism or grating to isolate the light of a particular wavelength, a device for holding the sample, and a photoelectric cell for measuring the intensity of the light. Your laboratory instructor will explain the operation of the spectrophotometers in your laboratory.

If your spectrophotometer is a Spectronic 20 with a meter, a commonly used instrument in general chemistry laboratories, a word of advice is offered here. The meter on this spectrophotometer is calibrated linearly in percent transmittance (100 T) and logarithmically in absorbance. You will always want to obtain three significant figures in your measurements of absorbance. However, if the absorbance is relatively large, doing so will be difficult on this meter because of the logarithmic scale. Percent transmittance can always be read with high precision, however, if it is above 10%. If you cannot measure an absorbance with three significant figures, measure the percent transmittance with this precision, convert to transmittance by dividing by 100, and calculate the absorbance by taking the negative logarithm of the transmittance.

Beer's Law

Beer's law states that the absorbance is directly related to the concentration (c) of the substance that absorbs light, or

$$A = kc$$

where k is a constant. Because A is a dimensionless number [$A = -\log_{10}(I/I_0)$] and the unit of measurement for c is mol/L (M), it follows that the unit of measurement for k is L/mol (M^{-1}). As you will see shortly, "k" is only a constant for a given substance at a particular wavelength. Its value may be zero if no light is absorbed at a particular wavelength, or it may be as high as $10^4\ M^{-1}$.

An Absorption Spectrum

Suppose the absorbance of a colored substance in a colorless liquid is measured at each of a series of wavelengths. Some typical results are given in Table C.1, where the absorbance was measured at intervals of 25 nm between 300 nm and 575 nm. These absorbances are plotted against the wavelengths in Figure C.1. After the data are plotted, the points are connected by a *smooth* curve. This curve, which represents the best estimate of the absorbance anywhere between 300 nm and 575 nm, is called an *absorption spectrum.*

Table C.1
Absorbances Obtained at Various Wavelengths for a Solution of a Hypothetical Substance ($c = 0.0120$ *M*)

λ (nm)	A	λ (nm)	A
300	0.002	450	0.558
325	0.016	475	0.281
350	0.144	500	0.092
375	0.341	525	0.031
400	0.578	550	0.004
425	0.681	575	0.001

FIGURE C.1

A typical absorption spectrum. The experimental data plotted here are given in Table C.1.

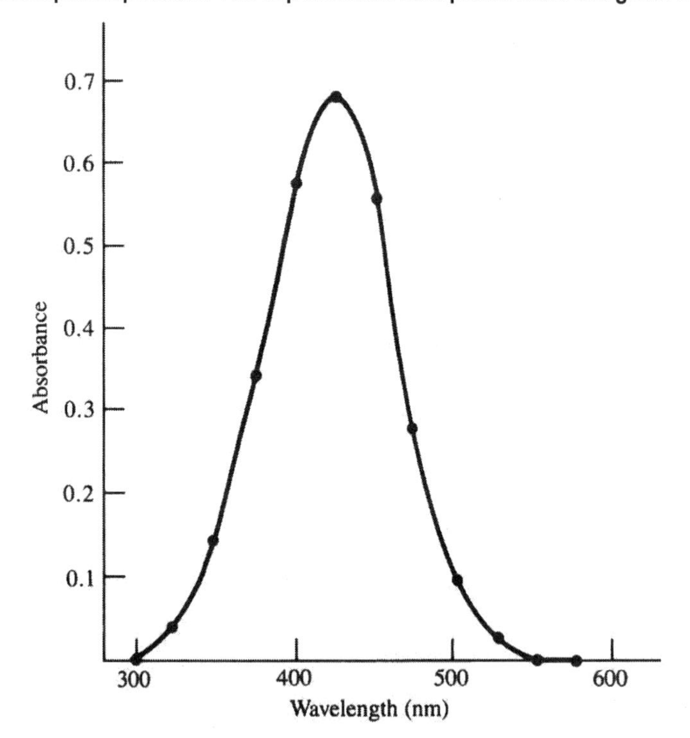

Because the concentration of the substance is fixed in this experiment, the change in the absorbance that is shown in Figure C.1 indicates the manner in which k from Beer's law is dependent on the wavelength. The value of the constant reaches a maximum at 425 nm, where the absorption spectrum for this substance reaches a maximum.

The Determination of k

The equation for Beer's law, $A = kc$, has the same form as the equation for a straight line, $y = mx + b$. A comparison of these equations indicates that

$$y = A$$
$$x = c$$
$$m = \text{slope} = k$$
$$b = \text{intercept on } y \text{ axis} = 0$$

Consequently, you should obtain a straight line when you plot the absorbances obtained at various concentrations against those concentrations. Moreover, the slope of that line will be given by k, and the line must pass through the origin ($A = 0$, $c = 0$), because the intercept is zero.

Let's use the data in Table C.2 as an example. You may assume that the hypothetical substance whose absorption spectrum is shown in Figure C.1 was used again to obtain these data. Figure C.2 shows a graph in which the absorbances in Table C.2 are plotted against the concentrations. A straight line passing through the origin was drawn in an attempt to provide the best fit for all the data. Experimental error is the reason why some of the points deviate from that line.

Table C.2 Absorbances Obtained at Various Concentrations at 425 nm

c (M)	A
0.0120	0.681
0.00960	0.540
0.00720	0.389
0.00480	0.270
0.00240	0.133

Any arbitrary point on the line will provide enough information for us to calculate the slope k. For example, we will choose a point with $c = 0.0100\ M$ and $A = 0.557$. The slope is given by

$$k = \frac{A}{c} = \frac{0.557}{0.0100\ M} = 55.7\ M^{-1}$$

FIGURE C.2
An example of Beer's law, $A = kc$. The slope of the straight line is given by k. The data were taken from Table C.2.

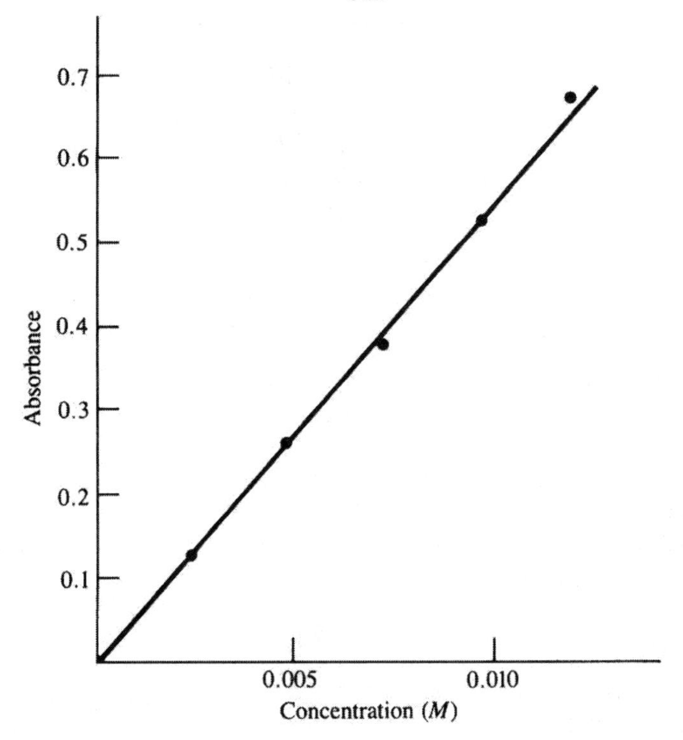

Linear Regression—A Better Way

How do we know whether Figure C.2 shows the best straight line? If five different people used their eyes to draw what they considered to be the best straight line, five slightly different straight lines with five slightly different slopes would undoubtedly result. Clearly, visual fitting of data to a straight line is not entirely satisfactory.

Fortunately, linear regression analysis (also called *the method of least squares*) provides the means to find an entirely reproducible straight line. If five people treat the data in Table C.2 by this method, the result will be the same straight line.

To begin, we know that the best straight line must pass through the origin. Although this constraint is not necessary, it simplifies the arithmetic and lessens the tedium of the calculations. The equation for this line will be $y = mx$, and the slope m will be given by

$$m = \frac{\Sigma xy}{\Sigma x^2}$$

The quantity $\sum xy$ in the numerator is the sum of the products of each x and y, and $\sum x2$ in the denominator is the sum of the squares of each x. In terms of the equation for Beer's law, k becomes

$$k = \frac{\Sigma cA}{\Sigma c^2}$$

The data in Table C.2 are subjected to linear regression analysis in Table C.3. Note that the k obtained in this manner differs slightly from the one derived from a visually fitted straight line. As a result, the best procedure is to use linear regression to calculate the slope and then to draw a straight line with that slope.

Table C.3

An Analysis by Linear Regression

c (M)	A	cA (M)	c^2 (M^2)
0.0120	0.681	0.0081720	0.0001440
0.00960	0.540	0.0051840	0.0000922
0.00720	0.389	0.0028008	0.0000518
0.00480	0.270	0.0012960	0.0000230
0.00240	0.133	0.0003192	0.0000058
		Sum = 0.0177720	Sum = 0.0003168

$$k = \frac{\Sigma cA}{\Sigma c^2} = \frac{0.0177720\,M}{0.0003168\,M^2} = 56.1\,M^{-1}$$

Appendix D: Indicators, pH Paper, and pH Meters

The pH of a solution can be either estimated, using acid–base indicators or pH paper, or measured directly with a pH meter. Each of these methods will be discussed in turn.

Indicators

Acid–base indicators (Ebbing/Gammon, Chapter 15) are usually very complicated molecules that are intensely colored. These substances are also weak acids or bases. For that reason, many aspects of the chemistry of these indicators are very similar to those of other weak acids or bases.

Methyl orange, for example, is an acid–base indicator and a weak acid. Because of its complicated nature, we write its formula in abbreviated form as HIn. This substance dissociates partially in solution according to

$$HIn(aq) + H_2O(l) \rightleftharpoons H_3O^+(aq) + In^-(aq)$$

The color of the acid form of this indicator is red, whereas the color of In^- is yellow.
The position of the equilibrium between HIn and In^- depends on the pH of the solution to which the indicator has been added. According to Le Châtelier's principle (Ebbing/Gammon, Chapter 14), a large concentration of $H3O^+$ ions (low pH) will cause the equilibrium to shift almost completely to the left. The color of the solution will then be red, the color of HIn. At lower concentrations of $H3O^+$ ions (higher pH), the equilibrium will shift from left to right, resulting in various hues of orange. If the equilibrium is shifted almost completely to the right, the color of the solution will be yellow, the color of In^-.

It is important to note that because an indicator is intensely colored, only small amounts are required. Because only small amounts are used, the indicator does not measurably alter the pH of the solution to be tested.

Consider the chart in Figure D.1, which gives the colors of four indicators as a function of pH. Returning to methyl orange, you will see from the chart that a solution containing this indicator will be red if the pH is less than 3.1, orange if the pH lies between 3.1 and 4.5, and yellow if the pH is greater than 4.5. By using all of the indicators given in the chart, you should be able to estimate an unknown pH that lies between 1.2 and 9.6. If the pH does not lie within this range, you will at least be able to say that it is less than 1.2 or greater than 9.6.

Because we must depend on the color of an indicator to estimate the pH, the solution to be tested must be colorless or very nearly colorless. This is one of the principal drawbacks to the use of indicators.

FIGURE D.1

Color changes for four acid–base indicators. Shaded areas indicate the pH intervals in which the colors change.

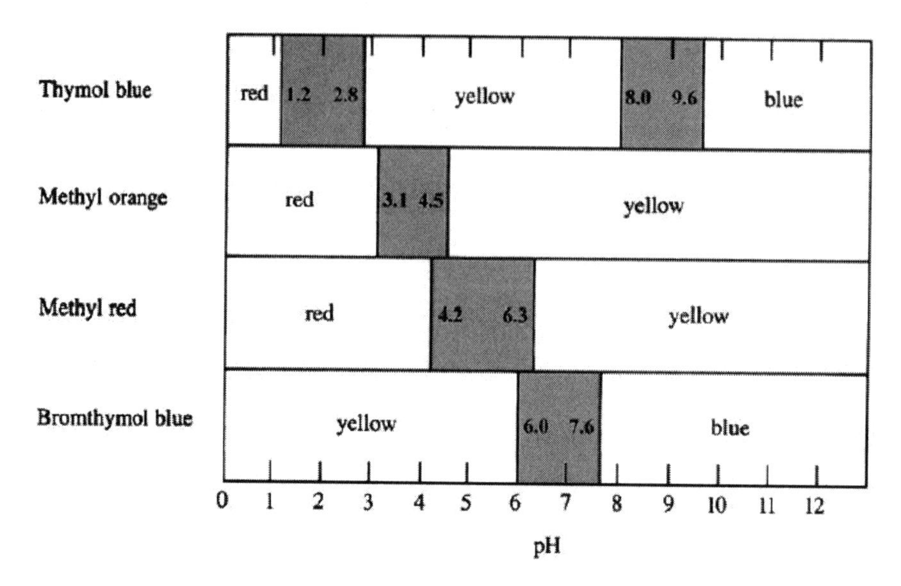

pH Paper

Paper strips that have been treated with a mixture of indicators can be used to estimate the pH of a solution. The indicators are chosen such that each one will change color at a different pH. The pH is estimated by moistening the paper with the solution being tested and then matching its color with a color on a chart provided by the manufacturer of the paper. The strips of paper are called *pH paper.* Again, colored solutions cannot be used.

Both wide-range and short-range papers are available. A wide-range paper might cover eight to eleven pH units (say, pH 1 to 11), whereas a short-range paper might cover only two or three pH units (say, pH
1 to 2.5).

pH Meters

A pH meter and its electrodes form a sensitive electrochemical device that makes possible the accurate, reproducible, and reliable measurement of the pH of a solution. Moreover, the solution does not need to be colorless.

There are several exterior designs for commercial pH meters. Meters may appear to differ from one another because they come from different manufacturers or, if they come from the same manufacturer, because they are different models with different prices. The differences usually lie in the way the measured pH is displayed, the positions of the control knobs, the types of electrodes, and the manner in which these electrodes are held in position.

Any pH meter, no matter how it looks, is just a voltmeter that measures the voltage of an electric current flowing through a solution between two electrodes. There is a direct relationship between the voltage and the pH of the solution (Ebbing/Gammon, Section 19.7). As a result, the meter on the instrument is calibrated directly in pH units rather than in volts.

Two electrodes are required. One of them is called a *glass or indicator electrode.* This electrode is sensitive to the concentration of H_3O^+ ions in the solution. The other is called the *reference electrode.* Its operation is virtually independent of the composition of the solution. These electrodes are sometimes combined into a single entity called a *combination electrode.* However, there are really two different electrodes present.

Although the operating rules for a pH meter depend on the model and the manufacturer, there are several steps you will need to follow with any instrument.

1. The electrodes should always be kept in a solution except when you are transferring them from one solution to another. When you transfer them, avoid contaminating the solutions. During the transfer, rinse the electrodes with a stream of distilled water, and catch the water in a beaker. Remove the excess water from the electrodes with tissue paper before you immerse the electrodes in the next solution. Do not touch the electrodes with your hand. Handle them with care because they are fragile.
2. If there is a knob that adjusts the pH meter for different temperatures, it should be set to the temperature of the solution whose pH is to be measured. More often than not, this temperature is also the temperature of the laboratory.
3. The pH meter must be calibrated or standardized with a solution whose pH is known before you can measure an unknown pH with accuracy. These solutions of known pH are called *buffer solutions*
4. Place your solution in the smallest container that is consistent with the experiment. Under the simplest of circumstances, you can measure the pH of a few milliliters of a solution in a large test tube with a combination electrode or in a 50-mL beaker with two electrodes.
5. You should be able to read the pH of a solution about 10 s after the electrodes have been immersed. The reading should be steady and not abruptly changing. If sudden changes do occur, consult your laboratory instructor. Additional operating rules may be issued by your laboratory instructor.